THE FRONTIERS COLLECTION

THE FRONTIERS COLLECTION

The books in this collection are devoted to challenging and open problems at the forefront of modern science, including related philosophical debates. In contrast to typical research monographs, however, they strive to present their topics in a manner accessible also to scientifically literate non-specialists wishing to gain insight into the deeper implications and fascinating questions involved. Taken as a whole, the series reflects the need for a fundamental and interdisciplinary approach to modern science. Furthermore, it is intended to encourage active scientists in all areas to ponder over important and perhaps controversial issues beyond their own speciality. Extending from quantum physics and relativity to entropy, consciousness and complex systems—the Frontiers Collection will inspire readers to push back the frontiers of their own knowledge.

More information about this series at http://www.springer.com/series/5342

For a full list of published titles, please see back of book or springer.com/series/5342

Hans J. Pirner

THE UNKNOWN AS AN ENGINE FOR SCIENCE

An Essay on the Definite and the Indefinite

Springer

Hans J. Pirner
Institut für Theoretische Physik
Universität Heidelberg
Heidelberg
Germany

Translated by William D. Brewer

ISSN 1612-3018 ISSN 2197-6619 (electronic)
THE FRONTIERS COLLECTION
ISBN 978-3-319-38637-9 ISBN 978-3-319-18509-5 (eBook)
DOI 10.1007/978-3-319-18509-5

Springer Cham Heidelberg New York Dordrecht London

Softcover reprint of the hardcover 1st edition 2015
Translation from the German language edition: *Das Bestimmte und das Unbestimmte* by Hans J. Pirner,

Printed on acid-free paper

Springer International Publishing AG Switzerland is part of Springer Science+Business Media
(www.springer.com)

We have now not only traveled through the Land of Pure Reason and gained a view of all of its parts, but we have also surveyed it and determined each of its entities in their respective places. This land, however, is an island, and it has been enclosed within immutable boundaries by Nature herself. It is the Land of Truth (a delightful name), surrounded by a wide and stormy ocean, the veritable locus of illusion, where behind many a fog bank and many an iceberg, soon to melt away, new regions peep out; and where the roaming mariner, in search of new discoveries, is continually deceived by empty hopes and enmeshed in adventures from which he can never escape, but yet will never bring to a conclusion.

Immanuel Kant, *Critique of Pure Reason*

Prolog

At the first meeting of our group in 2009, our mentor Wolfgang Schluchter explained his views on interdisciplinary cooperation in the *Marsilius Kolleg*.[1] He reminded us that interdisciplinary work has something to do with the discipline of participating every week in our common meetings and with respecting the limits of our own knowledge. In addition, in his remarks he praised the achievements of the Western '*universitas*,' the community of scholars and the scholarly disciplines, which had brought us together in the *Kolleg*.

At that time, the project that led to this treatise did not yet exist, but when Andreas Kemmerling established a group to deal with the topic of 'vagueness,' I joined it. Within this group, I gave a lecture on the subject of the 'Indefinite.' I began by collecting various examples of indefiniteness in physics, which are summarized in the second chapter of this book. I stayed near my own area of expertise, since it had become clear that I could not catch up on the deficits of my knowledge of other fields. The original topic of 'vagueness' has survived in this text as a term for the semantic indefiniteness of language. In the course of a year, Bernd Schneidmüller joined the effort with his examples taken from medieval history; I refer to them in treating indefiniteness in Sect. 4.5, 'Following the Signs.'

There is a considerable body of scholarly research on the topic of 'the indefinite.' This concept itself is vague, so that one can imagine a variety of subtopics subsumed within it. The paradoxical result of successful research would be to eliminate the indefinite altogether. The investigation of this concept is thus accompanied by a certain risk, and there is no simple result in sight.

What is the correct key for approaching the indefinite? From the history of European music, we know how harmonies are revealed, then hidden, and once again conjured up. All the notes are equal, and the abrasive dissonances arise from a

[1]The *Marsilius Kolleg* at the University of Heidelberg is an organization whose goal is to encourage discussion and cooperation between the various academic cultures. It focuses on bridging the gap between the physical and biological sciences on the one hand, and the humanities, the social sciences, and jurisprudence on the other.

pact with the devil, as described by Thomas Mann in his '*Doktor Faustus*',[2] 'Hear the dying chord—even in its dispersed condition, it stands for a technical totality which contradicts reality. Every note supports the whole, the whole story in itself.' The indefinite is just such a dissonance for those toiling in the fields of scholarship. It is not as unpleasant as off-key notes, but is just as unloved.

Information, a modern version of the definite, can act as an antidote to the indefinite, an antidote which, however, in too strong a dosage only amplifies our disorientation; this phenomenon is all too familiar in our modern world. We have to deal with the many sources of information and evaluate their worth. Computers can process information, but they cannot think. Computer programs must be written by human beings.

Dealing with the indefinite demands a special approach. I will show how the definite and the defining are mixed in with the indefinite. The fragmentation of our knowledge creates many boundaries between the islands of the known and the ocean of the unknown which surrounds them, so that we can easily lose track of their interrelations, while we are navigating among them. In order to encompass reality, we need to see the overall picture: the known and the unknown exist together in our world.

Our knowledge is dispersed among the individual disciplines, which have necessarily specialized and have branched out enormously. Interdisciplinary cooperations can build bridges between those specialized fields. I owe my thanks to my colleagues in the *Marsilius Kolleg*, whose support made this study possible.

This English translation of *Das Unbestimmte und das Bestimmte* has been updated and revised in order to take into account scientific progress since the first German edition.

Heidelberg, 2015

[2]Thomas Mann: *Doctor Faustus*; see e.g., http://www.dartmouth.edu/~german/German7/Faustus.pdf.

Acknowledgments

I would like to thank my colleagues M. Cederbaum, J. Eichberger, B. Falkenburg, D. Fine, H.L. Harney, J. Hüfner, H. Horner, and C. Hungar, and in particular the members of the Marsilius Kolleg, A. Kemmerling, H.-G. Kräusslich, B. Schneidemüller, and W. Schluchter, for reading parts or the whole of this treatise and for helping me with their critical comments to improve the text. Thanks are due to J. Halfwassen for introducing me to the philosophy of Plato. Mr. T. Just from the collegium assisted in the publication of the treatise.

I also wish to thank my wife, Heide-Marie Lauterer, for serving as lector for the final version. I am grateful to W.D. Brewer for the careful translation of *Das Unbestimmte und das Bestimmte* into the present English version with the title *The Unknown as an Engine of Science*, which has been made possible by partial support from the Marsilius Kolleg.

Acknowledgments

Contents

List of Figures

List of Tables

Chapter 1
Introduction

Our knowledge is increasing at a great rate; the number of scientific and scholarly publications per year has doubled in the past 20 years, but at the same time, the boundaries of our knowledge are also becoming more and more extended. In many-dimensional systems, for example for a sphere of n dimensions ($n \gg 3$), it is well known that the major portion of its volume lies near the surface. If we represent our knowledge as the volume of a sphere and its indefiniteness as the surface, we can see that the indefinite increases just as rapidly as does our overall knowledge. Modern humans have to live with the indefinite. Our society turns to science with questions whose answers lie hidden in darkness. The hope is that science can shed light on the answers. Can these expectations be fulfilled? How does science deal with the indefinite?

1.1 When Does Science Become Indefinite?

This essay intends to investigate where the sciences meet up with the indefinite, with uncertainties and vagueness, and how they deal with them. Scientists and scholars attempt to escape from indefinite situations by making new assignments and re-ordering their old knowledge. There is always the hope that a Theory of Everything (TOE) will be formulated which can explain the fundamental components of the universe and their elementary interactions. It is a great challenge to unite the Standard Model of elementary particle theory with the theory of gravitation, thus obtaining a unified theory which could explain all the known phenomena with fewer parameters. Physicists also play with indeterminate mathematical models in order to extend the limits of our knowledge. They dispense with the conventional formalism and try to 'escape' into the not yet determined, thereby gaining openness and more freedom of choice. Since methods for solving nonlinear problems with varying length and time scales have improved, complex systems can be increasingly well described by mathematical models. These models, it is hoped, will also permit reliable predictions of complex relationships.

H.J. Pirner, *The Unknown as an Engine for Science*,
The Frontiers Collection, DOI 10.1007/978-3-319-18509-5_1

To me, it seems that the humanities have more experience in dealing with the indefinite. For example, an historian can invent a story based on as-yet indeterminate sources which plausibly connects individual events. Historical sources are seldom unambiguous, and thus are subject to differing interpretations. As a rule, this indeterminacy is taken account of in scholarly research into historical facts. But occasionally, is not one or another opinion forgotten in the flow of historical research? When we as individuals forget something, our memories become indistinct. Psychologists have recognized that forgetting permits new ideas to form which would have been precluded by precise memories. Although modern brain research has been able to localize memory functions using imaging techniques, the processes of memory are still not well understood.

Uncertainty is often the result of indefiniteness. Our view of the future appears cloudy when we do not understand the past. "It is very difficult to make predictions, especially when they concern the future".[1] This apparently paradoxical statement characterizes the empirically motivated sciences, which construct models based on past experience in order to extrapolate them into the future. But no one is certain whether such models contain all the relevant parameters, and it is unclear as to whether new events will occur in the period to which the predictions apply, that will influence the predictions; i.e. it is uncertain whether such models can deliver reliable forecasts of future events. Uncertainty is a component of many analyses of complex processes. Risk assessments are intended to guarantee that our decisions do not lead to catastrophes—but the probabilities of unlikely events occurring are highly uncertain. When such unexpected events in addition give rise to serious consequences—one can think for example of the nuclear catastrophe in Fukushima following the tsunami in 2011—then the futurologists find themselves confronted by a serious dilemma. Nassim Nicholas Taleb associates such events with the image of the "black swan".[2] If one has seen only white swans, the expectation is that no black swans exist. But they *do* exist; in Australia for example, black swans are indeed found in large numbers. Taleb uses this example to illustrate the unforeseen development of the Internet and the financial crises in the years 2000 and 2008. Can we learn from the experience of the stock-market traders that the indefinite is important, for correctly evaluating scientific results as well?

The fears arising from uncertainty increase the pressure to elucidate the indefinite. An indefinite diagnosis of an illness increases the fears of many people that the worst possible case is at hand. These psychological aspects of the indefinite cannot be ignored, even when we are discussing our knowledge of the universe or the accuracy of our conceptions of the world. They need not be considered simultaneously, but they have to be kept in mind in the background.

---

[1]This quotation has been attributed to various authors, among others Karl Valentin, Mark Twain and Niels Bohr.

[2]Taleb [1].

1.2 Questions About Uncertainty

Why should we consider the indefinite and the definite together? Cannot we be content with what we know for certain? When we are planning a research project, we begin by summarizing what has already been worked out in the field of interest. Only then do we move to the unknown and new in order to formulate our research goals. This departure into the new is supported by the fact that we take along a case packed with established tools, methods and results of past work. Based on our knowledge of the field, we can speculate as to where something new remains to be discovered. But are there additional categories of the unknown, and what can we learn from this (negative) catalog? Is the indefinite composed of individual singular points like black holes, which remain hidden from our knowledge, or does it form contiguous regions, like the boundaries of knowledge? Is there a horizon beyond which we cannot see, no matter how we intensify our research efforts?

Where is the indefinite to be found? Does chance play a role? Why are predictions of the future questionable? Where is the boundary between the known and the unknown? Is it a clearly-defined border? Can we localize the known, the definite "within" this boundary, and the unknown, the indefinite, "outside" it? If we define a system as a number of things working together as a complex ordered whole, does that system constitute something definite and its environment the indefinite? These questions cannot be answered in an abstract fashion; they require concrete examples. Can those examples be classified? Can one distinguish between trivial, superficial indefiniteness and the profoundly indefinite? Or better yet, between the indefinite which can be eliminated, and that which is an essential property of the problem at hand?

Assuming that there is a well-defined boundary between the indefinite and the definite, can that boundary be shifted? How does the boundary move when more information becomes available? What can be determined through information? How does an exchange of information between the indefinite environment and the definite system take place, if we designate the two separate regions in this way? Is "complex" just another name for "indefinite"? Or are there structures in complex systems? Can we better recognize them if we consider the indefinite and the definite together? What reasons are there for considering them together? Is this merely a question of knowledge? Or does "definite" simply mean naming things, so that we can more clearly distinguish them?

These abstract questions need to be answered in a concrete way, so that it becomes clear that the indefinite and the definite form a unity—which is however often not perceived as unified, since there seems to be no reason to consider them together.

1.3 Indefinite or Definite?

Since the word 'indefinite' or 'not definite' contains a negation, we are tempted to define it '*ex negativo*'. But its subconcepts overlap and are not mutually exclusive. I will distinguish six different meanings: random, uncertain, indeterminate, vague, indistinct, and undefined.

Something is indefinite when it occurs *randomly*, i.e. when it is not determined in advance. Randomness is based on a subjective lack of knowledge, characterized by a special situation, a missing context, or insufficient information. It arises through subjective ignorance and disappears when the subject obtains more information.

If the concept of "indefinite" refers to predictions of future events, one speaks of *uncertain*. There are for example uncertain prognoses. Our knowledge of future events is in general unreliable. But even careful measurements yield results which scatter around some mean value; they are subject to inevitable measurement errors and are in this sense uncertain as predictions of a precise value.

This should be distinguished from the *indeterminacy* of events which is dictated by nature itself and is not due to our lack of knowledge. In quantum mechanics, the Heisenberg uncertainty relations hold; they are best referred to as 'indeterminacy relations'. Quantum phenomena are intrinsically not sharply determined, not just because they are perturbed by our observation of them. Something which is not fixed by existing laws can also be referred to as indeterminate; its goals are unclear.

Language expressions are indefinite in the sense of *vague* when they cannot be objectively determined to be either true or false. I therefore associate the term *vagueness* with semantic indefiniteness. Vague concepts such as "rich", "small", or "warm" can indeed be given a numerical value in terms of an amount of money, a length or a temperature; but this value is arbitrary. Our everyday speech is full of such vague expressions. The statement "This room is warm" does not define the exact temperature of the room. Many of our instructions for actions contain vague expressions. "If the room is too warm, please turn down the heat" is a vague directive. Popper expresses his skepticism towards the possibility of exact definitions: "The false ("essentialist") doctrine that we can define (or "explicate") a word or an expression, that we can "fix" its meaning or make it "precise", corresponds in every respect to the incorrect doctrine that we can prove or secure or justify the truth of a theory".[3] Is vagueness simply a problem of our language? Are there perhaps limits in the empirical world leading to vague concepts of which we are still unaware?

Knowledge or memories are *indistinct* when the associated names, concepts or places are unclear. In dreams and imaginings, our experiences and fantasies are blended into an indefinite mixture. A map drawn on a very large scale provides no information about the locations of small places. Its lack of clarity or *indistinctness* is

---

[3]Popper [2].

due both to the large distances and to our own inability to distinguish detailed differences.

A further meaning of the concept of the indefinite is *undefined.* An object or a situation is undefined when it cannot be precisely distinguished from others. Theoretically, the lack of clear boundaries corresponds to undefined sets. Counting procedures can lead to this type of indefiniteness: Imagine that in a counting process, certain criteria are used to choose a subset. Then there can be elements which are neither a part of this set nor are outside it; they are located in a border region. Such a region often occurs in quantitative assessments. Consider the following example: At the beginning of a semester, new students apply for admission to a university. Some of them fulfil the admission criteria and some do not. In addition, there are a certain number of students who represent borderline cases and cannot be clearly sorted into the one category or the other.

What is one to do with the students who do not clearly belong to either of the categories? They can be grouped together, or perhaps preferably treated individually, as undefined borderline cases. The ambivalence towards foreigners in a society, who belong to it but at the same time stand outside it, is an interesting sociological phenomenon. Some aspects of this question have been investigated by Zigmunt Baumann.[4]

There is more information about the concept of "the definite" than about the concept of "the indefinite". The '*Philosophisches Wörterbuch*' defines a concept as *definite* when it is distinguishable from other concepts in terms of content and scope.[5]

The "*Cambridge Dictionary of Philosophy*" defines (completely) determined and determinable as "The color blue, e.g., is a determinate with respect of the determinable color; there is no property F independent of color such that a color is blue if and only if it is F. In contrast, there is a property, having equal sides, such that a rectangle is a square if and only if it has this property. Square is a properly differentiated species of the genus rectangle".[6]

A similar meaning of definite is "precise" or "uniquely determined". Charles Sanders Pierce distinguishes between definite and determined: "Accurate writers have apparently made a distinction between the *definite* and the *determinate*. A subject is determinate in respect to any character which inheres in it or is universally and affirmatively predicated of it".[7] We would say that something is determined with respect to an intrinsic, characteristic property. A general symbol which can stand for various different specifications and is not in any respect determined is called *vague* by Pierce. In the aspects in which a general symbol is not vague, he calls it "definite". He assigns the different concepts of "definiteness, generality and

[4]Bauman [3].

[5]*Philosophisches Wörterbuch*, founded by Heinrich Schmidt, extended by Georgi Schischkoff, Stuttgart 1961.

[6]Audi [4].

[7]Pierce [5].

Table 1.1 Various meanings of the concepts of "the indefinite" and "the definite", with examples to illustrate them

Indefinite	Definite	Example
Random, chance	Probable, determined	Statistical events
Uncertain	Reliable; open	Predictions
Indeterminate	Sharp, well-defined	Quantum objects
Vague	Unambiguous	Language
Indistinct	Clearcut	Memory
Undefined	Defined	Elements of sets

vagueness" to a category which was associated by Immanuel Kant with the *quantity of judgments*.[8] Kant's *quantity* can best be understood in terms of a counting procedure which, as explained above, leads to undefined special cases.

The converse of the indefinite is likewise plurivalent. Thus the opposite of random is probable. In predicting the future, a prognosis may be uncertain, but it may also be found to be reliable, and in any case can be regarded positively as being 'open' for different outcomes. Table 1.1 sets out the different meanings of the "indefinite" and "definite" which we have discussed above. A skepticism due to the indefinite can be neutralized by the manifold possibilities for determining the definite. A balanced treatment of a topic requires that the indefinite and the definite be considered together.

1.4 How to Proceed

The phenomena of indefiniteness and definiteness in various fields have their own importance in that they indicate different causes of the indefinite. In the following investigations, I will restrict myself to the six definitions of the indefinite and the definite given in Sect. 1.3, and will search for examples for these categories from the sciences. In choosing the examples in Chap. 2, I have followed my own interests and will concentrate in particular on those types of indefiniteness that are familiar to me as a physicist. However, I have made an effort to situate this list within a general framework which includes other scientific disciplines, as well.

In Chap. 3, I have attempted to describe how indefiniteness can be resolved by information. Starting from the mathematical definition of information, I investigate various types of information. I maintain that information becomes all the more valuable for a system the more it increases the system's complexity and reduces the indefiniteness of its surroundings. The definition of a measure for the indefiniteness allows to quantify the value of the information.

The definite system and its indefinite surroundings can exchange information. Niklas Luhmann has elaborated how our increasingly differentiated and highly

[8]Immanuel [6].

complex society is pushing back its boundaries. In the end, it extends out into the indefinite: "The more complex a system is intended to be, the more abstractly must its boundaries be defined".[9] In order to arrive at a unified system, indefinite and indeterminable complexity must be excluded, while the internal complexity is being ordered and managed. Networks play an important role in this process, since they are multidimensional.

Chapter 4 is devoted to the question of how the indefinite can be determined; how can one deal reasonably with indefiniteness? I will discuss various methods of determining the indefinite. Experiments and observations carried out with increasingly refined instruments yield ever more precise results which extend the borders of knowledge. Decisions made under uncertain conditions can be prepared with the help of theory. The rules of fuzzy logic permit vague technical instructions to be carried out. A preliminary stage of the formulation of theories can be considered to be the transfer of indefinite data into a known environment. Language makes use of metaphors in order to focus a topic sharply. Theories themselves contain symbols or tokens in which the indefinite and the definite are mixed. If more than one interpretation is possible, scenarios which offer alternatives become necessary. I will explain these concepts using examples in order to illustrate how they operate in practice.

This essay is written from the viewpoint of a physicist. It is directed at scientists and interested laypersons from all fields who are curious about exploring the limits of our knowledge. This work does not attempt to speculate about future theories, although an effort was made to include the latest research results in it. Right at the forefront of research is the uncertainty greatest. I deal here with the indefinite and speak to all those who deal with indefiniteness on a daily basis: doctors who are struggling to make the right diagnosis, legal practitioners who have to deal with the ambiguities of laws, or computer scientists who are searching through their data for significance. In order to keep the scientific discussion at an understandable level, I have explained the technical terms in a glossary. Complex and technical physics discussions are indicated in the text by smaller print and can be skipped over in a first reading. In treating this topic, I wish to make the offer of initiating a dialogue between various cultures of knowledge; therefore, I have permitted myself to bring different disciplines into the discussion, although I am not able to present their viewpoints in full. The various cultures of knowledge can most readily meet at the forefront of the not-yet-known when they are in the process of developing and unfolding their specific methods.

Areas of research which are still indefinite are best suited for this, since they do not impede new ideas through the territorial claims of particular disciplines or historically developed argumentation.

---

[9]Luhmann and Habermas [7].

References

1. Taleb, N.N.: The Black Swan—The Impact of the Highly Improbable. London (2007)
2. Popper, K.: Vermutungen und Widerlegungen – Das Wachstum der wissenschaftlichen Erkenntnis (English original edition: Conjectures and Refutations, The Growth of Scientific Knowledge, London 1963), p. 615. Tübingen (2009)
3. Bauman, Z.: Moderne und Ambivalenz – Das Ende der Eindeutigkeit (English original edition: Modernity und Ambivalence). Hamburg (1992)
4. Audi, R. (ed.): The Cambridge Dictionary of Philosophy, p. 228. Cambridge (1999). See e.g. https://archive.org/stream/RobertiAudi_The.Cambridge.Dictionary.of.Philosophy/Robert.Audi_The.Cambridge.Dictionary.of.Philosophy#page/n0/mode/2up
5. Peirce, C.S.: Issues of pragmaticism. In: Houser, N. (ed.) The Essential Peirce. Selected Philosophical Writings, vol. 2, p. 352. Bloomington, Illinois (1998)
6. Immanuel, K.: Kritik der reinen Vernunft. In: Gross, F. (ed.) §9 Von der logischen Funktion des Verstandes in Urteilen, p. 98, Leipzig (1920). English: see e.g. http://staffweb.hkbu.edu.hk/ppp/cpr/toc.html
7. Luhmann, N., Habermas, J.: Systemtheoretische Argumentationen – Eine Entgegnung auf Jürgen Habermas. In: Theorie der Gesellschaft oder Sozialtechnologie, p. 301. Frankfurt am Main (1971). English: see e.g. http://ecpr.eu/Filestore/PaperProposal/30ae4fa5−2413−4e4d−a837−37c5df56f4c7.pdf

Chapter 2
Evidence for the Indefinite

2.1 Chance in Theoretical Models

We begin by examining the type of theoretical indefiniteness in physics which is due to our partial, subjective lack of knowledge of the properties of individual particles within large systems. Historically, this problem was first treated in the latter part of the 19th century. At that time, thermodynamics was enjoying great success; it had developed empirically from the treatment of heat engines and the study of chemical reactions. It could successfully describe macroscopic phenomena, for example the rise in pressure in a steam boiler when its temperature is increased. Nevertheless, Ludwig Boltzmann developed his own theory—*statistical mechanics*—which refers to microscopic particles (atoms or molecules) that had not been directly observed at the time and whose dynamics were unknown. In one mole of oxygen, i.e. 16 grams, there are more than 10^{23} molecules; 10^{23} is a *very* large number, a 1 followed by 23 zeroes. Each molecule has three spatial coordinates and velocities in each of the three spatial dimensions. The amount of data required to specify the positions and velocities of all of those molecules makes the task of determining them seem hopeless.

In order to attack this problem, Boltzmann introduced chance into physics. He started with microstates whose details are random, determined by chance; they form a statistical ensemble, which is determined by only a few macroscopic state variables. Microstates are chance elements of a larger whole, the macrostate, which is well defined. The macrostate is described by physical quantities such as temperature, volume and the number of particles, which lie in the realm of classical measurement methods. Boltzmann applied the concept of "entropy", originally introduced by Robert Clausius and Nicolas Léonard Sadi Carnot. It is a measure of our lack of knowledge of a system—the indefiniteness of the state, so to speak. The less we know about the microstates, i.e. the less information we have about the system, the greater its entropy.

H.J. Pirner, *The Unknown as an Engine for Science*,
The Frontiers Collection, DOI 10.1007/978-3-319-18509-5_2

Boltzmann's opponents, the so-called "energeticists", maintained that all of the laws of thermodynamics could be derived from energy conservation alone. Boltzmann countered with the statement that the energy available to perform work is determined by the entropy. Energy is conserved in all processes, but the *quality* of the energy decreases in technical processes as the entropy increases. For example, we have little use for the exhaust gases of an automobile; their entropy is too high, so that they cannot be used to power engines or drive chemical processes. When the entropy of a system is high, its energy is worthless.

Mathematically, statistical mechanics is based on probability theory. Just as in a casino or a card game, chance "determines" the dynamics of the atoms or molecules in a boiler. At a given temperature, the energy of a given atom can be specified only in terms of a probability. The distribution of energies has its maximum at the average energy. Variations or fluctuations around this average do occur, but they are unimportant for the increase of the pressure in the boiler with increasing temperature. In equilibrium, each microstate has a well-defined probability which does not change over time.

The approach to equilibrium presumes systems whose equations of motion are not integrable. Integrable systems usually consist of only one or two particles. For example, the one-dimensional motions of a single point mass are determined solely by energy conservation, and the system is integrable. The orbital motion of a planet around the sun can be completely and analytically described by applying the conservation of energy and of angular momentum. In non-integrable, chaotic systems there are no additional conserved quantities besides energy and momentum which could predetermine the system's evolution. The ensemble of all the positions and velocities of the individual particles moves over a hypersurface which is determined by energy conservation. Such a system begins with a spatial distribution of the velocities and position coordinates that is sharply localized. If we wait long enough, this cell spreads out like a drop of blue ink in a glass of clear water. Such systems are termed "ergodic".

The question arises as to whether a statistical description can be applied only to simple objects such as atoms—can it also be applied to complex time-dependent phenomena such as biological or economic processes? In the year 1827, the English botanist Robert Brown used a microscope to observe a pollen grain suspended in a liquid. He noted that the grain was moving, as though it were alive. It took almost 80 years until Albert Einstein was able to explain this process in terms of a random walk.

Einstein made the assumption that the light molecules of the liquid transfer momentum to the heavier pollen grain by colliding with it, and cause it to recoil in a random way. The forces which act during these collisions are not determined in detail; we do not know their directions and strengths. On the whole, these forces average out to zero. They have values within a zone between $-F$ and $+F$ around zero. Nevertheless, the pollen grain moves. The reason for this is that within a certain time interval, the forces are correlated, i.e. they do not change abruptly from large negative values to large positive ones. Physicists parametrize these correlations. Einstein succeeded in solving the equation of motion of the heavy particle in the presence of the random external forces. The solution leads to a mean square velocity which at long times is proportional to the strength of the correlation function. If the particle comes into equilibrium with the liquid after some long time, then there is a simple relationship between the strength of the internal friction and the mean square random force.

Similar differential equations with random external terms can be used to describe complex biological or economic systems. The random "forces" might be political crises or climatic catastrophes which influence the target parameter, e.g. the price of a stock or of a commodity. In these generally very large systems of equations, a great number of interactions are parametrized, insofar as their parametrization is possible. The remaining influences which cannot be analyzed are random perturbations, of which one knows only mean values and correlations. With the goal of learning as much as possible about the world around us, physics has dealt with objects about which we will never obtain information with certainty. Nevertheless, using the combination of probability calculus and differential equations, it has been possible to reach a better understanding of such systems. The statistical treatment of medium-sized quantum systems raises new questions. Here, the indefiniteness in the description of individual systems combines with the intrinsic indeterminacy of quantum mechanics, which I will describe separately later. The number of particles in such systems is small compared to thermodynamic systems; we are dealing here with perhaps 100–200 particles. The systems are themselves small in size, i.e. they are subject to the laws governing quantum objects (cf. Sect. 2.3). The spacing of their energy levels is comparable with the resolution with which one can determine the levels. In contrast, in thermodynamic systems, the energy levels are so closely spaced that it would be hopeless to try to compute or measure them individually. In the mesoscopic (medium-sized) systems considered here, the mean level density and the behavior of the energy-level spacing are accessible to experiments. Examples of mesoscopic systems are atomic nuclei at excitation energies of a few million electron volts, or quantum billiards, i.e. two-dimensional racetracks of electrons on solid surfaces or at interfaces. The theoretical treatment of these systems models their indefiniteness in a unique way and can thus describe certain aspects of the energy levels.

The treatment of systems of this kind was initiated by Niels Bohr, who coined the term "compound nucleus" for the combined nucleus which is formed during the scattering of a nuclear projectile from an atomic 'target' nucleus. In low-energy collisions of neutrons from a reactor with atomic nuclei, the projectile chooses one of the many possible final states only after "circling" around the target nucleus for some time. This long contact time leads to a quasi-equilibrium state which permits a statistical description of the system. Since the work of Werner Karl Heisenberg, it has been known that wave mechanics gives an equivalent representation of every quantum-mechanical problem with a well-defined matrix (in Heisenberg's "matrix mechanics"). Modern methods are based on a theory which models a class of energy matrices, determined by symmetry alone, instead of a single quantum-mechanical energy matrix with sharply defined energy levels.

Physical theory cannot parametrize and solve the complex interactions of even these in principle relatively few particles. However, it can successfully model a statistical distribution of matrices which reflects the important symmetry properties of the system.

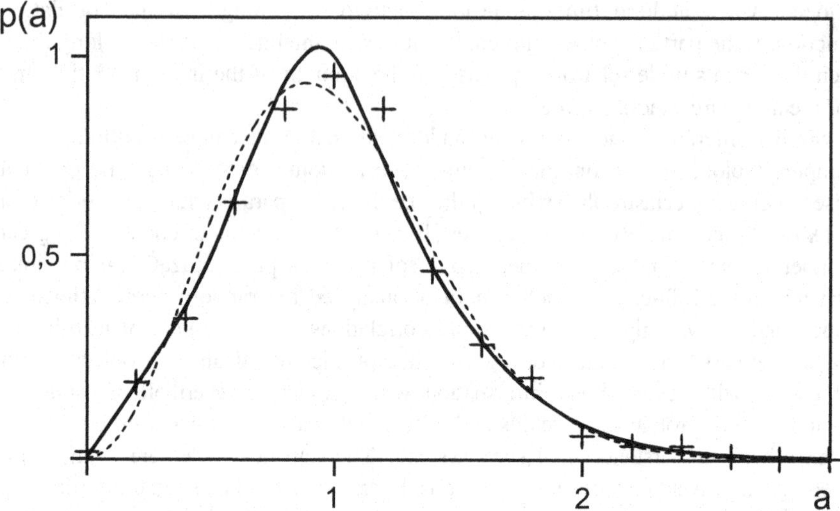

Fig. 2.1 The scaled distribution of the spacing of parked automobiles is plotted as a function of their bumper-to-bumper distance a for the automobiles on the right-hand side of the street; after Hradec Královė. The value $a = 1$ on the horizontal axis corresponds to the average distance. The *dashed curve* shows the distribution given by a Gaussian unitary random-matrix ensemble. The *solid curve* was calculated from a detailed model (Šeba [2]; see also earlier references by Abul-Magd [3]) which distinguishes between the left-hand and the right-hand sides of the street

One refers to these models as *random-matrix theories*.[1] Note how theoretical physicists introduce structure into the limiting cases that make up the twilight zone. Symmetry is necessary in order to limit the number of possible cases and to correlate the classes of matrices with classes of phenomena. This approach has proved successful in many areas of low-energy nuclear physics, quantum chaos, number theory, and the physics of disordered solids.

Random matrices can also successfully describe the spacing between parked cars. Gaussian unitary or orthogonal random matrices lead to a distribution of the distances from bumper to bumper which has a simple form (cf. Fig. 2.1). One finds a characteristic distribution which begins at zero, passes through a maximum and then drops slowly again towards zero. There are only a few virtuoso drivers who manage to park very close to the car ahead. I have to add that the empirical studies were not carried out in Paris, where it is well known that the cars behind or ahead of a space are often shoved together by desperate motorists, leading to a spacing which makes it impossible for pedestrians to cross the street.

The statistical method has even attracted proponents who assert that the fundamental form and the coupling constants in the Standard Model of elementary-particle physics are the result of a stochastic averaging over generically known interactions.

[1]Bohigas and Weidenmüller [1]. The authors give an introduction to this relatively new area of physics.

This would mean that the question of a deeper understanding of the 40 unknown couplings of all the elementary particles and the form of their interactions could be attributed to a random model. Holger Bech Nielsen,[2] an exponent of this random dynamics, starts with the assumption that the truely fundamental elementary-particle theory is extremely complicated. He presumes that, independently of this complex fundamental theory, at the highest energies effective laws have emerged whose form is determined randomly.[3] This yields an approach to elementary-particle physics which has not attracted many supporters. The majority of physicists are in fact convinced that the fundamental laws of physics become simpler and simpler as one investigates smaller and smaller elementary objects. This assumption seems well justified if one considers how physics has evolved towards smaller and smaller size scales. From the physics of molecules, atoms, atomic nuclei, and nucleons to the physics of quarks, one cannot deny a certain trend towards simplification.

There is an almost metaphysical debate between the proponents of a rationally-planned world and the supporters of the idea that chance played a significant role even in the origin of the universe. Many of the exponents of these random cosmologies believe that only the anthropic principle can be made responsible for choosing our universe out of the multitude of possible existing universes ("the multiverse"). This principle states that our universe has precisely those properties that are necessary so that we can live in it; the theory thus derives the laws governing the origin of the world from observations of its current state.

In string theory, the indefiniteness is based not on our lack of knowledge of the elementary excitations, but rather on the uncertainty about the background environment in which the strings move. There are presumed to be 10^{500} possible realizations ("landscapes") of this background, which give rise to a similar number of vacua. One can imagine the background to be like a potential landscape (a contour map), with hills and valleys in which the fields or strings seek their lowest energy states like balls rolling on a hilly surface. Among these enormously many possible states is the one state which corresponds to our universe, in which the elementary electric charge, the velocity of light and the mass of the proton have precisely their known values in our world. In cosmology, our universe plays the role of just one variant within a 'multi'verse. The other universes cannot be observed, and thus their existence is rather controversial in modern theoretical physics.

The opponents of string theory and the multiverse insist that most of the possible string theories are not consistent with the observations of type IIa supernovas,

---

[2]Nielsen and Brene [4].

[3]Nielsen and Brene, *ibid.*: "In the search for the most fundamental theory of physics, one usually looks for a simplest possible model, but could it not be that the fundamental "World Machinery" (or theory) could be extremely complicated? We see that we have some very beautiful and simple laws of nature such as Newton's laws, Hooke's law, the Standard Model and so on—how could such transparency and simplicity arise from a very complex world? The Random Dynamics project is based on the idea that all known laws of nature can, in a similar way as Hooke's law, be derived in some limit(s), practically independent of the underlying theory of the World Machinery. The limit which could suggestively be the relevant limit for many laws would be that the fundamental energy scale is very large compared to the energies of the elementary particles, even in very high energy experiments. A likely fundamental energy scale would be the Planck energy."

which document an accelerating expansion of the universe.[4] Those theories predict with a high probability exactly the converse behavior. A simplified account of recent understanding of a small positive cosmological constant is given by Raphael Bousso,[5] and Bousso together with Joseph Polchinski.[6] They argue that such stable vacuum solutions are possible in string theory and are selected in the cosmic evolution by us as observers (anthropic principle).

Pierre Duhem, in his "Theory of Physics", described such a confused situation as a preliminary state of affairs: "In any case, a state of indecision never persists for long. The day arrives when common sense points clearly to a particular theory, and the opposing fractions give up their resistance, although there is no purely logical reason to do so".[7] The experimental physicists are presently hoping that the new LHC (Large Hadron Collider) will put an end to the confusion. Whether it will in fact support the speculations of string theory or will contradict them remains to be decided. Results of the first experimental run of the LHC have not given any evidence for physics beyond the standard model.

String theory has proved to be such a fruitful branch of mathematical physics that it will continue to be investigated, possibly in complete disregard of the experimental results. Duhem's assessment of theoretical physics may prove not to be correct.

Indefinite in the sense of *random*

Examples

- The positions and velocities of individual atoms in a well-defined gas;
- The forces on a pollen grain suspended in a liquid;
- The matrices and the spacing of parked automobiles;
- Our universe within the multiverse.

2.2 The Unpredictable Future

Dante Alighieri[8] in the twentieth canto of his "Divine Comedy" describes how he meets up with the fortune-tellers in Hell: "Their faces were turned backwards, and they had to move their feet in a backwards direction; they no longer had any way of

[4]Ellis and Smolin [5].

[5]Bousso [6].

[6]Bousso and Polchinski [7].

[7]Duhem [8].

[8]Dante Alighieri: *The Divine Comedy*; see for example http://www.gutenberg.org/files/8800/8800-h/8800-h.htm.

seeing ahead". Dante lets the fortune-tellers walk with slow-moving steps through the twisted valley in Hell. Just as the fortune-tellers are punished for their false prognoses, incorrect predictions can decisively influence the lives of individuals or of society. In order to elucidate the uncertain future, one must analyze the past and investigate the possible future with the aid of models. We shall ask in the following sections just how uncertain predictions of the future based on such models really are.

A physical model represents a simplified picture. We cannot see light waves, but nonetheless we use a wave concept to describe light, since it exhibits phenomena similar to water waves. Models are thus heuristic representations which describe new phenomena. A theory is more detailed than a model and therefore has a higher status. It evolves on the basis of many phenomena which it sums up with the aid of physical laws, such as energy conservation; these are concepts that provide relationships between the fundamental quantities. Physicists have laws at their disposal in classical mechanics which predict the position and velocity of an object at a later time from knowledge of its position and velocity at one particular time, presuming that they know the forces which act upon it. A point like object (a "point mass") follows a path that satisfies Newton's equation, which predicts precisely its position and velocity at a later time $t > t_0$ from their known values at the earlier time t_0. This is also possible for several point masses with complex interactions. The French mathematician and astronomer Pierre-Simon Laplace (1749–1827) was the first to prove the constancy of the mean orbits of the planets in celestial mechanics. The solar system is thereby so stable that it will "remain intact" up to the final inflation of the sun some 10^9 years from now. Classical mechanics is a deterministic theory of the world; it computes the future unambiguously from knowledge of the past.

Our predictive powers are to be sure rather limited when we consider systems with many degrees of freedom. The solutions of the equations of motion of many-body systems depend so sensitively on their initial conditions that even a minimal deviation in those initial values leads to a large variation in the predicted results after a finite time. One terms such systems "chaotic", since the predictions may change erratically due to small variations in the initial conditions. Since we can never determine the initial values with absolute precision, our ability to predict events precisely diminishes rapidly with increasing time. If one improves the precision of the initial values for a chaotic system by a factor of 100, then the length of time for which predictions hold within a given margin of error is increased by only a factor of 4. The components of the system and their interactions are themselves well-defined and determined, but their future evolution is uncertain, since the exact initial state of the system can never be precisely known.

Similar problems occur when we model clouds, winds and air currents in order to predict our weather and climate. The meteorologist Edward Norton Lorenz[9] was the first to point out the chaotic solutions of the hydrodynamic equations, which are functions of time and space. He distinguishes two types of predictions: Predictions

[9]Lorenz [9].

of the first kind refer to the time evolution of a system as a function of its initial conditions in time with given boundary conditions in space. Predictions of the second kind refer to the solution of the equations with fixed initial conditions but variable spatial boundary conditions. Weather predictions are of the first type, since a finite number of satellites and weather stations cannot completely and precisely determine the initial weather conditions. They depend upon the local boundary conditions in the atmosphere and in the oceans.

It is a simple truth that long-term predictions, e.g. over a time period of 20 years, would have to contain the technological and political changes that will occur during those years. As we well know, it is not sufficient simply to extrapolate observed trends from the past, since here real discontinuities and qualitative improvements or setbacks can occur. Climate predictions refer to long periods of time, while weather prognoses apply to the short term. For the weather, a time frame of 5 days is a considerable challenge; for the climate, one has to consider time periods of several years or even decades. The preferred method in the latter case is to model the atmospheric circulation only in large time steps and then to relate it to the dynamics of the oceans, the biosphere, the polar regions and other slowly-changing components of the climate. Relatively good results have been obtained for the El Niño events, a cold circulation in the Pacific which appears every three to six years.[10] At the beginning of the year 2009, the National Weather Service in the USA predicted a temperature increase of 2 ± 0.5 °C for December 2009, which could affect the yield of fisheries.[11] Of most concern is the possibility that instabilities could arise in the long-term climate as a result of human activities, for example an interruption of the flow of the Gulf Stream in the North Atlantic,[12] or global warming due to CO_2 emissions.

Model calculations in the form of equations of motion or partial differential hydrodynamic equations are imprecise due to chaotic dynamics, amplified by the insufficiently-known initial conditions. Statistical predictions, which become necessary when we know too little about the system itself, should be distinguished from such calculations. There, we make use of probabilities to characterize the system. If within a statistical description we assume that the parameters of the system do not change within the time period covered by the forecasts, then the path is clear for authoritative predictions. We will give two examples of the application and the limitations of statistics.

In Germany, the average height of the whole population is 178 cm, with a variance of 7 cm. This means that within a probability of 70 %, the height of a German citizen lies between 171 and 185 cm (178 ± 7 cm). The height distributions of adult men and women in Germany are shown in Fig. 2.2. The bell-shaped curves show that women on the average are somewhat shorter than men.

---

[10]http://www.cpc.noaa.gov/products/.

[11]This prediction was rather well fulfilled at the time of writing. For the week centered on December 23, 2009, the temperature anomaly reached 1.94 °C, as quoted in https://bobtisdale.wordpress.com/2009/12/.

[12]Hasselmann [10].

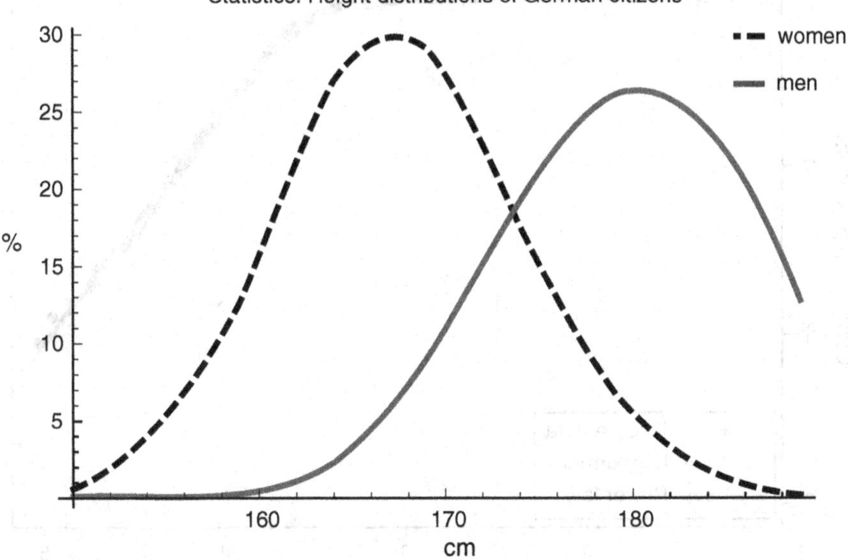

Fig. 2.2 The probability distribution [Figure from http://de.wikipedia.org/wiki/K%C3%B6rpergr%C3%B6%C3%9Fe, dated 02.06.2010; *Source* statista.org, Socio-economic Panel (SOEP).] of the heights of men and women in Germany. They resemble bell curves, whose maxima coincide with the average height of men and of women

Physicists refer to the corresponding probability distributions as *normal distributions*. Nassim Nicholas Taleb attributes them to a fictitious country called "Mediocristan", the land of mediocrity, where everything occurs "normally".

If I have an appointment with a business partner whom I have not met previously, I can say with a high probability that he or she will be between 171 and 185 cm in height. If I know whether the partner is a man or a woman, I can make this prediction more precise, since the average heights and their variances are different for men and women. The problem becomes more difficult if I attempt to estimate the economic status of my unknown partner, as manifested in his/her income or net worth. In practice, this estimate will have to be made from the social context; it makes a big difference whether one is meeting with a sales representative or with the CEO of a corporation.

Let us now assume that we have an appointment with a CEO, and thus with a presumably wealthy person. Taleb places a group of such people in another country, which he calls "Extremistan". In Extremistan, different laws hold from those of Mediocristan, i.e. the probability distributions there do not have the bell-curve or normal-curve shape that we saw for the height distribution within the general population; instead, they exhibit power laws like shown in the income distribution of the wealthiest (Fig. 2.3). The Italian economist Vilfredo Pareto[13] discovered such

[13]Pareto [11].

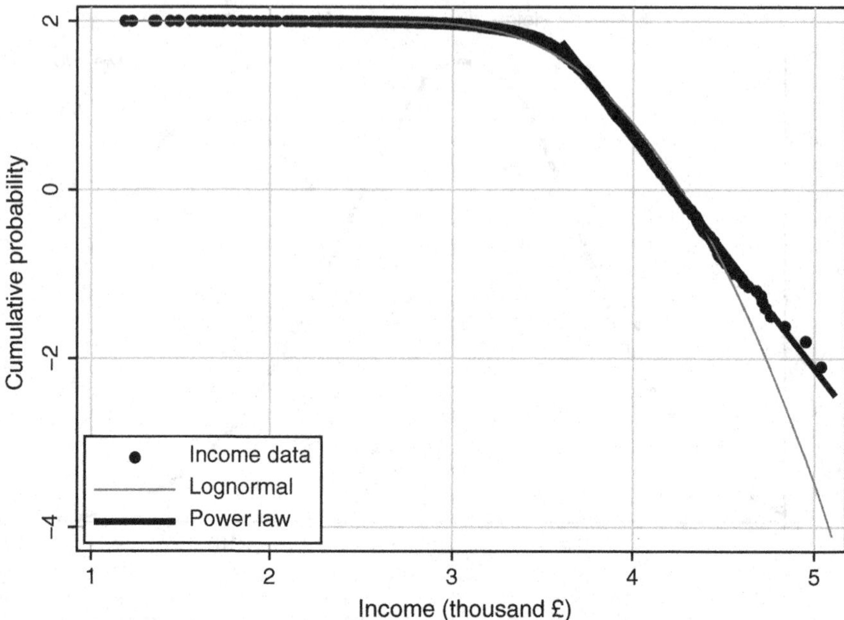

Fig. 2.3 The cumulative, i.e. integrated income distribution (Clementi und Gallegati, *ibid.*) of Italian households in the year 1998. The *curve* shows which percentage of the population (the number 2 corresponds to $10^2 = 100$ %) has an income higher than the income on the horizontal scale (the number 3 corresponds to $10^{3+3} = 1{,}000{,}000$ Lire). Up to very high incomes, the data are fit well by a log-normal distribution (*thin line*). For the "very wealthy", they follow a power distribution, which is higher in comparison to the log-normal distribution (*full line*, "fat tail")

distributions empirically at the end of the 19th century. The probability of encountering great wealth is given by the distribution $P(w) = ak^a/w^{1+a}$ with the index $a = 1.4$ and the minimum parameter k, if one accepts the analyses of Moshe Levy and Sorin Solomon,[14] or of Fabio Clementi and Mauro Gallegati.[15] The distributions of price variations, the sizes of corporations, fluctuations of stock and bond prices and their returns also follow similar power laws. In particular, investment returns have the same index as the distribution of wealth, which led the above authors to the hypothesis that the richest segment of the population invests large amounts of money and then profits the most from the returns on those investments.

Physicists are fascinated by scaling laws of the above type, because they occur in many systems near critical points. The critical point of a magnetic material is defined as the temperature at which it loses its macroscopic magnetization. Below this point, long-range spatial correlations between the elementary magnets or magnetic moments occur, which obey power laws. This process is to be sure quite different from economic processes.

---

[14]Levy and Solomon [12].

[15]Clementi and Gallegati [13].

In economic terms, success has no natural upper limit. As in the game of chance "Winner takes all", or in the Biblical saying, "He that hath, to him shall be given", wealth can increase enormously; it grows without limits, in contrast to the height of individuals, which is subject to a natural upper limit.

What does all this have to do with the reliability of predictions of future events? For the following discussion, we shall assume a power-law distribution and ignore the deviations of the distribution of wealth from the power law at low and middle asset values (they follow a log-normal distribution there). Then the mean value of wealth is indeed well defined by the power-law distribution, but in the case of an index of $a < 2$, the variance from the mean is not. We found the value $a = 1.4$; it is thus difficult owing to the broad fluctuations to determine the limits of the amount of assets of our business partner. Likewise, it may prove impossible to predict the future behavior of other quantities which obey scaling laws, when their variances are undefined.

The question "What is meant by 'wealthy'?" is thus not vague in the sense that there is no sharp boundary between rich and poor, as is asserted by some philosophers. The answer to this question is adequately provided by the empirical distribution of wealth. There is a well-defined point at which the log-normal distribution that applies to average citizens departs from the power law that describes the assets of the wealthy.

The indefiniteness in the case of power laws results from their undefined variances. In Extremistan, we would have difficulties in making predictions about the future. The economic crises and collapses of 1987, 1997, 2003, and 2008 would seem to verify this statement. As recently as 2002, Harry Eugene Stanley and Rosario Nunzio Mantegna[16] optimistically discussed the successes of *econophysics*: "In fact, power law distributions lack a typical scale, which is reflected by the property that the variance is infinite for $a < 2$. One important accomplishment is the almost complete consensus concerning the finiteness of the variance of price changes (on the stock market)". They were convinced that the fluctuations of stock-market prices are calculable, and on the basis of this hypothesis, they defended the efficiency of the market, which supposedly adjusts prices quickly and rationally to their optimum values. The term "econophysics" was coined by these authors and describes the interdisciplinary analysis of empirical economic data using physical models. More than in axiomatic-mathematically-oriented macroeconomics, in this new area of physics an attempt is made to keep research closely connected to the data, which elucidate both distinctive structural characteristics and also invariants through international comparisons.[17] The French theoretical physicist Jean-Philippe Bouchaud[18] pointed out that predictions based on log-normal distributions "with finite variances" by Fischer Black and Myron Scholes were in part responsible for the crisis in 1987, and he affirmed the need for more pragmatic and realistic models.

[16]Mantegna and Stanley [14].

[17]Sinha et al. [15].

[18]Bouchaud [16].

The normal distributions miss out on the paradigm of the black swan,[19] i.e. the occurrence of rare but significant events.

The prediction of truly time-dependent processes is additionally complicated by the variation of their parameters. Bouchaud is of the opinion that the agents of the system (e.g. the stock market) impair the predictability of economic systems through their irrational actions. Most investigations of economic systems have emphasized the financial markets; only in more recent times has there been modeling which takes into account both economic *and* ecological aspects. This can provide a more realistic, if still unfocused picture of the future, since it considers new interactions, but also brings additional uncertainties into play.

Indefinite, in the sense of *uncertain*.

Examples

- Chaotic systems;
- Weather and climate predictions;
- Distributions of wealth with "fat tails";
- Long-term prognoses, owing to technical or political changes.

2.3 Indeterminacy in Quantum Physics

Physics, as an empirical science, is thought to be the essence of definiteness. It is based upon experiments and observations which can be repeated and reproduced at any time and in every place with the aid of suitable experimental conditions. Ideally, an experiment yields measured values which form the basis for hypotheses. Hypotheses which have been verified by a number of experiments can be merged into natural laws. The basic facts underlying physics originate in Nature itself. They are independent of the models which we use to describe Nature. Are there nevertheless examples of indefiniteness within physics? The answer is "yes"!

One can detect uncertainty in the results of measurements which ought to yield identical values when they are carried out by different groups of experimenters. Even when the experimental conditions are controlled with great care, every measurement apparatus will yield a somewhat different value for each individual measurement. Uncontrollable external influences such as small temperature variations, fluctuations in voltage or motions of the air can lead to errors in a measured value. Measurements are always associated with uncertainties. Measurement errors of this type (*statistical errors*) can be treated using statistical methods as described

[19]Nassim Nicholas Taleb, see footnote 3 in Chap. 1.

in Sects. 2.1 and 2.2. The specific "uncertainty" in quantum physics is particularly fascinating, since it is of a quite different nature. I will deal with this difference in the following.

The basis of our discussion are the indeterminacy relations due to Werner Karl Heisenberg, one of which states that the position and the momentum (momentum = mass times velocity) of a particle cannot be determined simultaneously with unlimited precision. The indeterminacy Δx in the measurement of the particle's position and the indeterminacy Δp in the measurement of its momentum together obey the inequality

$$\Delta x \times \Delta p > h/(4\pi)$$

where h is Planck's constant (also known as Planck's *quantum of action* by analogy to classical mechanics). Max Planck discovered this fundamental physical constant in 1900 through his observation that radiation of a particular wavelength can be emitted or absorbed only in discrete amounts (quanta) of energy. In the indeterminacy relation, Planck's constant limits the precision of a simultaneous measurement of the position and the velocity of a particle, independently of the potential accuracy of the measurement instrumentation. The more precise the determination of the momentary position of the particle, the greater is the indeterminacy *in principle* in its velocity determination. This relation corresponds to an indefiniteness which is inherent in Nature itself and does not depend on the quality of our apparatus or on the observer; therefore one should preferably speak of the "*indeterminacy relation*", since it is an objective fact and we have reserved the concept of "uncertainty" to describe a lack of subjective knowledge.

In quantum mechanics, the status of a particle is described by a function which treats it as a superposition of spatially-localized states. This "wavefunction" Ψ reflects a "virtual" state of the particle, which becomes real with the probability $|\Psi(r)|^2$ when the position of the particle is determined by a measurement at the location r. This probability is objective; it is not based on a subjective lack of knowledge on the part of the observer, as might play a role e.g. in determining the momenta of the individual atoms within a hot gas. The observer of a gas cannot know the velocities and positions of all of its constituent atoms, since it is simply not possible in practice to measure all of them.

The phrase "'virtual' state of the wavefunction" expresses the fact that the wavefunction itself cannot be directly observed. The absolute square $|\Psi(r)|^2$ of the wavefunction is however an observable quantity, and it gives the probability with which the particle is to be found at the position r. The wave nature of a quantum object gives rise to the indeterminacy relations. Broader, spread-out wavefunctions are associated with a greater uncertainty upon measurement.

In a velocity or momentum measurement, it is more expedient to deal with the virtual state within the space of all the possible momenta, i.e. the wavefunction in momentum space, $|\Psi(p)|^2$. The real-space and momentum-space wavefunctions are related to each other through an exact mathematical transformation, which was developed by Jean Baptiste Joseph Fourier. It is useful also for example in the analysis of the frequency spectrum of an acoustic signal. The broader the wavefunction in coordinate space, the sharper it will be in momentum space, and vice versa, as a result of this transformation property.

The indefiniteness in quantum mechanics can be understood by considering the following example: If one passes an electron through a double slit (e.g. a metal screen with two narrow, parallel openings), then one cannot say with certainty through which of the two slits it has passed, nor exactly where it will impact on a detector screen. The quantum-mechanical result for the impacts on the detector screen (cf. Fig. 2.4) corresponds to an intensity pattern as would be expected from *waves* which had passed through the two slits. The image reminds us of the superposition of two circular waves emitted from the centers of the two slits. We can produce and superpose circular waves for example by dropping two stones simultaneously into a calm lake; the distance between the points where they hit the water corresponds to the spacing of the two slits. Wave maxima from one wave which meet up with minima from the other cancel each other out, while two maxima or two minima reinforce each other when they meet. Mathematically, the *phase* of the waves defines the positions where at a given time a maximum or a minimum will be observed. The interference pattern which results from such an experiment is not simply the sum of the individual wave patterns from a single stone or a single slit.

In order to proceed with our discussion, we first need to deal with the probability statements $w(X)$ made by quantum mechanics.

We denote the detection of an electron on the screen by 'A', the passage of an electron through the upper slit by 'B', and passage through the lower slit by 'not B'. In classical mechanics, one would assume that the probabilities of passage through the upper and the lower slits simply add, as is in fact observed in the experiment with bullets (see Fig. 2.4), i.e.

$$w(\mathrm{A}) = w(\mathrm{A} \cap \mathrm{B}) + w(\mathrm{A} \cap \neg \mathrm{B})$$

while in quantum mechanics, we find:

$$w(\mathrm{A}) = w(\mathrm{A} \cap \mathrm{B}) + w(\mathrm{A} \cap \neg \mathrm{B}) + \text{interference term}.$$

Peter Mittelstaedt[20] asserts that the new quantum-mechanical situation is a relative violation of the dictum "*tertium non datur*" of classical logic, since the deviation from two-valued logic is conveyed through the relevant probability. If the events where electrons passed through the upper or the lower slit corresponded to mutually exclusive sets, the formula given above would nevertheless contradict one of Kolmogorow's axioms of probability theory.

It is notable that the result of the quantum-mechanical double-slit experiment approaches the classical result when the vacuum within the apparatus is replaced by a gas that scatters the particles. The double-slit experiment was carried out by Klaus

[20]Mittelstaedt [17].

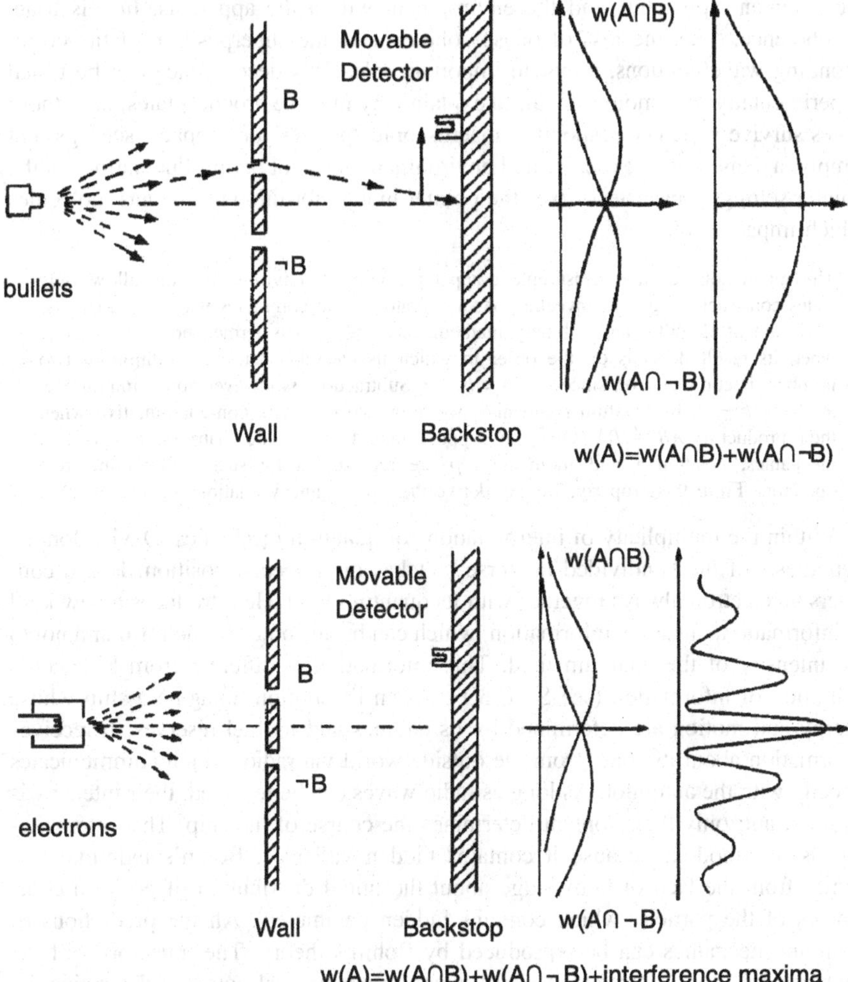

Fig. 2.4 The double-slit experiment: Bullets or electrons from a source (a gun or an electron gun) pass through two slit openings in a wall and impact onto a screen ("backstop"). The resulting pattern on the screen for bullets is simply the sum of the patterns obtained when only one of the two slits is open. In the case of electrons, interference maxima and minima are seen, which are due to the wave nature of matter (i.e. of quantum objects, here electrons)

Hornberger et al.[21] using Fullerenes, large spherical molecules composed of carbon atoms. The amount of gas within the apparatus determines the visibility of the interference pattern, which becomes fainter exponentially with increasing gas pressure. This experiment indicates that the measurement process must be explained in terms of the interactions between the system being observed, the

[21]Hornberger et al. [18].

measurement apparatus, and the environment within the apparatus. In this triad, decoherence,[22] i.e. the loss of phase coherence in the superposition of the corresponding wavefunctions, plays an important role. This decoherence can be tested experimentally. It is more difficult to explain why in these product states, only those states survive which correspond to macroscopic "pointer" or "impact" settings. Put simply, a dense gas as experimental environment does not permit fluctuations in the pointer settings, but instead fixes the pointer in equilibrium with the gas molecules which impact upon it.

The interactions of the measurement apparatus with its environment, i.e. allow only states constructed from eigenvectors of an operator which commutes with the apparatus-environment Hamiltonian. A binary mathematical operation is termed non-commutative when its result depends on the order in which its elements act. The addition of two numbers is commutative, since $a + b = b + a$. Subtraction is however non-commutative, $a - b \neq b - a$. In quantum mechanics, we term two operators non-commutative when their product is $AB \neq BA$. One can suppose that A and B are matrices. In quantum mechanics, position x and momentum p are represented by such non-commutative operators. From this property, we can derive the indeterminacy relations mathematically.

Within the multiplicity of interpretations of quantum mechanics, David Bohm's hypothesis of the "undivided universe"[23] takes on a special position. Bohm considers an electron always together with its quantum field. He introduces a new kind of information, an active information, which can be attributed to the form and not to the intensity of the quantum field. This information is different from Shannon's definition of information (see Sect. 3.1). As an illustration, imagine a ship whose energy and motion are determined by its engines and its fuel reserves. It receives information about its route from the outside world via radio, which communicates directly with the autopilot. As long as radio waves can be received, their intensity is unimportant; only their content determines the course of the ship. This interpretation is unorthodox, because it contains hidden variables. Bohm's indefiniteness comes from the lack of knowledge about the initial distribution of position coordinates of the particle, which contains hidden parameters. All the predictions of quantum mechanics can be reproduced by Bohm's theory. The trajectories of the coordinates within a somewhat weird quantum-mechanical potential determine the results of measurements.

There have been attempts to combine quantum mechanics and gravitation into a unified theory known as quantum gravity.[24] The combination of Planck's constant

[22]Joos and Zeh [19], Paz and Zurek [20].

[23]Bohm and Hiley [21].

[24]Parallel to these attempts, in the modern literature gravitation is often interpreted not as a fundamental theory, but rather as an emergent or effective theory, which holds only at length scales large compared with the Planck length λ. In that case, it would not be necessary to quantize gravity. The metric tensor would then be analogous to the density distribution of particles in thermodynamics. The new fundamental degrees of freedom are unknown. Indeed, the Einstein equations have a great similarity at a horizon to the thermodynamic equation $TdS - dE = PdV$. See e.g. Padmanabhan [22].

h from quantum mechanics, the speed of light c from special relativity, and the universal constant G of gravitation results in a length scale $\lambda = \sqrt{\hbar G/2\pi c^3} = 10^{-33}$ cm, below which quantum mechanics and gravitation can no longer be treated separately. In order to get an idea of the size of this *Planck length* λ, it is useful to define a logarithmic scale which represents shorter and shorter lengths by a factor of 10 at equal intervals. If we represent the size of a coffee cup by the first mark on this scale, we will have to place 13 marks along the scale to arrive at the size of an atomic nucleus. We would then have to continue three times as far along the scale of size reductions by a factor of 10 to arrive at the Planck length λ. Quantum gravity has evolved around a new indeterminacy relation,

$$\Delta x \times \Delta y > \lambda^2.$$

Here, the smallest elementary length, the Planck length, is on the right-hand side of the inequality. This indeterminacy relation would mean that it is not possible to measure a length more precisely than the Planck length. Every length measurement with increasing precision along the x direction would produce an increased indeterminacy in length measurements along the perpendicular y direction. In quantum mechanics, there is in fact an analogous behavior. The center of the circular orbit of an electron in a homogeneous magnetic field pointing along the z direction cannot be determined exactly.

The associated quantum-mechanical x and y coordinates are represented by operators whose product in this case is $xy \neq yx$; therefore, the center point of the electron's orbit cannot be determined with unlimited accuracy. The indeterminacy relation between x and y is at first sight surprising, since without the magnetic field, x and y are not complementary quantities in the sense of quantum mechanics, in contrast to the position x and the corresponding momentum p_x.

It follows from the indeterminacy relations of quantum mechanics that smaller and smaller details of the elementary particles can be investigated by using particle accelerators of higher and higher energies. These accelerators work like microscopes with a high resolution for matter waves.

This would no longer be the case if we were to reach the scale of the smallest elementary length. Leonard Susskind,[25] a resolute supporter of string theory, has extrapolated the hypotheses of that theory to collisions at the highest energies. While in the usual quantum mechanics, higher energies E allow us to study smaller and smaller length scales, $L = hc/(2\pi E)$, this behavior would change in such a string theory at extremely high energies.

In very high-energy collisions, a black hole would be created, which could emit only low-energy Hawking radiation whose energy would be of the order of the inverse Schwarzschild radius R. The Schwarzschild radius $R = 2GE/c^2$, which corresponds to the energy of the Hawking radiation, defines a horizon (the *event horizon*) behind which the black hole is hidden. The importance of gravitational effects enters through the gravitational constant. In

[25]Susskind and Lindesay [23].

astronomy, black holes are stars that have collapsed under the effect of the gravitational force; their gravitational potential is so strong that even light can no longer escape from the black hole. Therefore, such objects appear black in the sky. In principle, there could also be miniature versions of these black holes ('mini black holes'), which are created in collisions of particles at high energies.

There have even been legal proceedings which were aimed at preventing experiments with the new accelerators at the Brookhaven National Laboratory in the USA and at CERN in Geneva, because the plaintiffs feared that small black holes would be produced that could swallow up the Earth.

Experience to date with the similarly energetic cosmic radiation, along with the extremely short lifetimes of the hypothetical mini black holes, contradict the assumptions of those lawsuits. The experiments which have been carried out recently yield lower limits for the possible masses of these objects. No such small black holes have been observed. The Schwarzschild radius of the black holes defines the event horizon, behind which information appears to vanish. Correspondingly, black holes have no characteristic properties aside from their overall electric charge, their angular momentum and their mass. However, quantum mechanics requires that the state of the wavefunction at a given time determines the state of the wavefunction at every later time. One might say that the information contained in the wavefunction is conserved. Now, if one were to throw a book containing information into a black hole, the information would be annihilated, i.e. the above fundamental assumption of quantum mechanics would be violated. This contradiction[26] has recently been resolved through an improved understanding of the role of the Hawking radiation which is emitted by the black holes.

A horizon is also defined for us by the expansion of the universe. At the *cosmic horizon*, at a distance of 45 billion (4.5×10^{10}) light years from us, matter is moving away so fast that no light can reach us from further away. Behind this horizon the universe is inaccessible to us. The astounding homogeneity of the cosmic microwave background radiation can be explained by applying an inflationary model, i.e. an extremely rapid expansion at the beginning of the universe. Without this rapid expansion following the Big Bang, parts of the observable universe would not have been able to communicate and would have different temperatures, in contradiction to the observed homogeneity.

In the preceding summary of the indefiniteness in the quantum world, we have seen how the indeterminacy relations play an extremely important role in physics. They set limits to our possible knowledge, beyond which we can no longer peek at Nature's secrets. It was long thought that the intervention of the observer was the source of the indeterminacy or fuzziness of quantum measurements. Our modern understanding of quantum physics is however that these uncertainty relations must be attributed to Nature herself; they are *indeterminacy* relations. At the same time, these relations have become central, structuring elements of the associated theories.

[26]See Susskind and Lindesay, *ibid.*

Indefinite, in the sense of *indeterminate.*
Examples

- Simultaneous measurement of the position and the momentum of a quantum object;
- Impact of the quantum object at a particular location on a detector;
- Measurement of the position coordinates in different directions in quantum gravitation.

2.4 Vagueness in Language

In our exploration of the typology of indefiniteness, we meet up with phenomena whose origin is not to be found in Nature, but instead in language, in memory and in human thought. These are common in everyday life and have therefore been under discussion since ancient times. The same problems are topics of research today in modern linguistics, psychology, and philosophy. In the following account, I will describe both the ancient beginnings as well as the modern scholarly treatment of these questions. By their comparison, it can be seen how much effort has been necessary to achieve even a small amount of progress.

Linguistic indefiniteness[27] refers in the main to "discrete" sets, whose elements can be enumerated using whole numbers. A standard example is the question: "When is a certain man X bald?" In principle, this question can be answered by looking at X's head. But in making the assessment, one might have doubts as to just how few hairs justify applying the attribute "bald". As a result, the statement "X is bald" remains vague. The following question originated in ancient times: What is a sandpile? Is an accumulation of grains a pile? How many grains do we need in order to make a pile? This example gave rise to the Sorites paradox (from the ancient Greek σορὸσ = "pile"), which results from applying induction. The first step in the logical chain of reasoning states: An accumulation of *one* grain is *not* a sandpile. The second step asserts that if an accumulation of n grains is not a sandpile, then an accumulation of $n + 1$ grains is also not a sandpile. By induction, one then arrives at the paradoxical result that an accumulation of a thousand grains is still not a sandpile.

How have physicists dealt with the vagueness in the definition of a sandpile? A sandpile is an open system which is not in stable equilibrium and which continually exchanges energy and matter with its surroundings through drifting sand. Sand avalanches occur when the slope of the pile exceeds the angle of repose; then the pile abandons its metastable

[27]Kemmerling [24].

equilibrium and sand flows down its flanks. Per Bak, Chao Tang and Kurt Wiesenfeldder[28] have developed a simple model of how a sandpile can form. The basal area of the pile is divided into squares; in each square, there is room for up to K sand grains. Grains which are carried to the pile are randomly distributed over the squares. If a square contains more than K grains, the excess grains are moved onto adjoining squares. In this way, clusters are formed on the basal area, whose sizes are determined by the avalanches that occur within each square. The number $D(s)$ of the clusters as a function of their size s follows a power law, $D(s) \approx s^{-\tau}$, with $\tau = 0.98$. This distribution is independent of the quantity K, which could perhaps be compared to the size of the sandpile in the Sorites discussion. The pile's structure can be explained without having to fine-tune the parameter K; this is an advantage of Bak's model. As an aside, we note that in real experiments,[29] the critical behavior of the mathematical model given above cannot always be demonstrated. If e.g. rice grains are used, and they are too round, one observes deviations.

To give rise to the Sorites paradox, a set (e.g. of sandpiles) must be specified which cannot be broken up into smaller units. The induction step in the Sorites paradox leads us to believe that there are only gradual changes without qualitative differences. Taking as a counterexample the vague concept of childhood, we note that the transition from child to adult passes through puberty, which is accompanied by significant changes.

All of the discussions which have been carried on in philosophy seem to offer little practical guidance for finding a solution. I will now describe how engineers or computer scientists attack this problem: They interpret the continuous transition of one state into another as "fuzzyness".

In 1965, the systems analyst Lofti Asker Zadeh[30] introduced to this end the concept of "fuzzy sets". Elements belong to fuzzy sets only to a certain extent, not exclusively. A vague statement such as "this room is warm" does not define the temperature of the room, but it nevertheless often suffices to characterize the state of the room for its occupants. In order to describe the vague meaning of the word "warm" more precisely, it is useful to imagine an ensemble of warm rooms and then to compare the particular room under consideration with these rooms, to evaluate how appropriate or inappropriate the statement was. The foundation for this comparison is a range of values or *basis* for the problem at hand. In this example, the basis is a range of temperatures T, which for the purposes of this discussion and without loss of generality I will define as "warm" between 0 and 30 °C. This eliminates absurdities such as the Sorites paradox right from the beginning, namely that one could, by means of small reductions of the threshold for "warm" in steps of 1 °C followed by an induction step, arrive at the ridiculous idea that −1 °C must still be considered as warm. In the next step, we define a *membership function* for all of the elements t in T, which attributes a value between 0 and 1 to each element t. I will also call $m(t)$ with $0 < m < 1$ an "opinion function", since this term makes it clearer that we are dealing with a subjective assignment rather than an actual fixed meaning. This qualifies the problem of multi-valued logic right from the beginning. Opinions can be

---

[28] Bak et al. [25].

[29] Frette et al. [26].

[30] Zadeh [27].

inconsistent and need not always be logically justifiable. The value 1 of the opinion function, i.e. $m(t) = 1$, corresponds to the opinion that the statement is completely true, while $m(t) = 0$ corresponds to the opinion that the statement is totally false. The opinion function for the statement that A is *not* correct is defined as follows:

$$m(\neg \mathrm{A}) = 1 - m(\mathrm{A}).$$

The goal of this mathematical formulation is to quantify the indefiniteness of the judgements of experts. Note that this indefiniteness is quite different from the uncertainty which arises from the variation among different temperature measurements. The opinion function varies between 0 and 1, making it similar to a probability. It is however not normalized, and the indefiniteness that it denotes is not the same as the entropy or information derived from probabilities, as we shall see. We continue with our example by discussing various opinion functions (cf. Fig. 2.5). We first consider the "wishy-washy" expert (*ww*), who has no strong opinion and would therefore set the value of the opinion function to $m = ½$ over the whole range. In contrast, we have the "warm-blooded" expert (*w*), who finds the upper part of the range of temperatures to be quite warm and has a linearly increasing opinion curve. Our third expert is a "coddled" Middle European (*cw*) who finds only the range between 20 and 30 °C to be genuinely warm, and whose opinion function exhibits a sharp maximum at 25 °C. These various functions are shown in Fig. 2.5. We can then use information theory[31] to define a measure for the indefiniteness $U(\mathrm{A})$, which fulfils the condition that $U(\neg\mathrm{A}) = U(\mathrm{A})$,

$$U(\mathrm{A}) = -\int \mathrm{d}t\{m(t)\log[2, m(t)] + ((1 - m(t))\log[2, (1 - m(t))])\}.$$

Here, we use the logarithm to the basis 2; then we find for the indefiniteness of the three experts $U_{ww}(\mathrm{A}) = 30$ °C, $U_w(\mathrm{A}) = 21.6$ °C, and $U_{cw}(\mathrm{A}) = 7.2$ °C. This quantitative determination agrees with our impression that the *ww* expert had the least definite opinion, while the *cw* expert is rather sharply opinionated.

As the next step, I will define the opinion function for the simultaneous occurrence ("and") of the statements A and B by making use of the minimum function:

$$m(\mathrm{A} \cap \mathrm{B}) = \min[m(\mathrm{A}), m(\mathrm{B})].$$

The "or" conjunction can be similarly defined using the maximum function. These specifications allow us to quantify to what extent additional information would help to determine the state of the room as "warm".

Additional information could be a statement such as, "If children are wearing short pants" (B), and I assert that it is completely true ($m = 1$) if the temperature is above 18 °C.

[31]DeLuca and Termini [28]. See also: Bandemer and Näther [29].

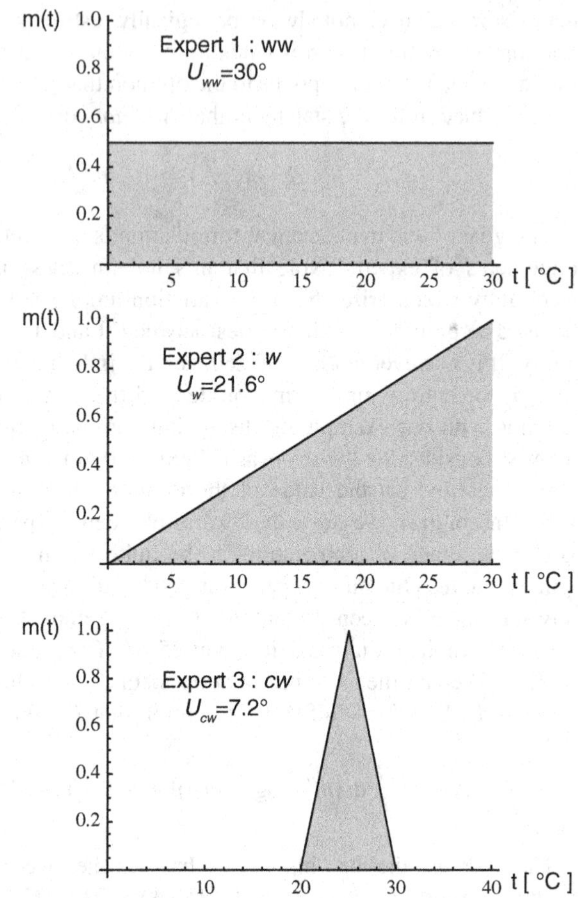

Fig. 2.5 Membership or opinion functions denote to what extent an expert considers the temperatures given on the horizontal axis to be 'warm'. From above, the opinion functions of the "wishy-washy" expert (*ww*), the "warm-blooded" expert (*w*) and the "coddled" expert (*cw*) are shown as functions of the temperature in degrees celsius. The values U_{ww} (A) = 30 °C, U_w (A) = 21.6 °C, and U_{cw} (A) = 7.2 °C give the indefiniteness of the opinions of the various experts in units of temperature, i.e. in degrees celsius; see the quantification of the indefiniteness (in the main text)

In Fig. 2.6, one must then take the minimum of the linear function and the rectangular function; the latter is equal to zero below 18°. Above 18°, this minimum coincides with the linear function.

This additional information reduces the indefiniteness of the opinions of the first two experts (*ww* and *w*), which were particularly vague. The indefiniteness of the opinion of the 'coddled' expert (*cw*) is not changed by the additional information.

Compare:

$$U_{ww}(\mathrm{A}) = 30\ °\mathrm{C},\ U_w(\mathrm{A}) = 21.6\ °\mathrm{C}\ \text{and}\ U_{cw}(\mathrm{A}) = 7.2\ °\mathrm{C}$$

with

$$U_{ww}(\mathrm{A} \cap \mathrm{B}) = 12.0\ °\mathrm{C},\ U_w(\mathrm{A} \cap \mathrm{B}) = 7.8\ °\mathrm{C}\ \text{and}\ U_{cw}(\mathrm{A} \cap \mathrm{B}) = 7.2\ °\mathrm{C}.$$

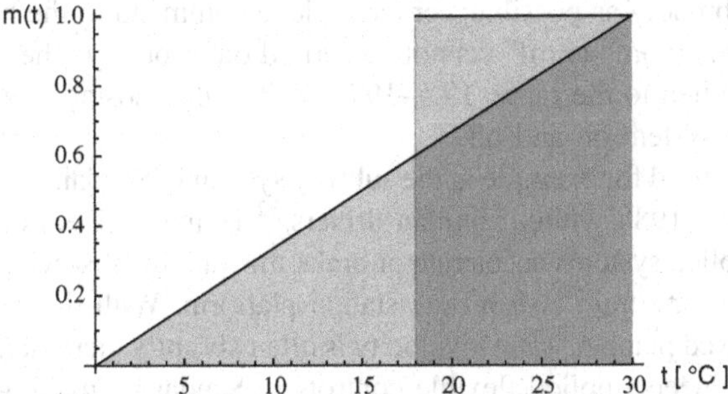

Fig. 2.6 Due to the conjunction (∩) of the two statements "the room is warm" and "the children are wearing short pants", we find new opinion functions $m(t) = 0$ for $T < 18$ °C and $m(t) = T/30$ °C for $T > 18$ °C (*darkly shaded area*). This leads to an indefiniteness of 7.8 °C for the "warm-blooded" expert (*w*)

The gain from an additional specification is clear for the first two experts; the second approaches the quantitative indefiniteness of the third when the new information is added.

I denote the opinion function which corresponds to the simultaneous occurrence of the statements "A" and "not A" as the *ambiguity*:

$$\textit{ambiguity}(\mathrm{A}\ \text{and}\ \neg\mathrm{A}) = m(\mathrm{A} \cap \neg\mathrm{A}).$$

This definition leads to useful results. In particular, for the opinion of the first expert (*ww*), we obtain a constant ambiguity of ½, while the ambiguity of the second expert (*w*) has a maximum at 15 °C (Fig. 2.7).

Thus it is most difficult for him at the temperature 15° to decide if the room is warm or not warm, i.e. to give a definite opinion. At the two endpoints of the range, i.e. at 0 and 30 °C, the situation is clear and the ambiguity vanishes.

The advantage of fuzzy set theory now consists in the fact that one can implement vague instructions on a machine. It is important for control technology to

Fig. 2.7 The ambiguity of the second expert (*w*) with respect to the statement "This room is warm" (*darkly shaded area*). His opinion is least strong where the two lines cross each other, i.e. the ambiguity is maximal at 15 °C

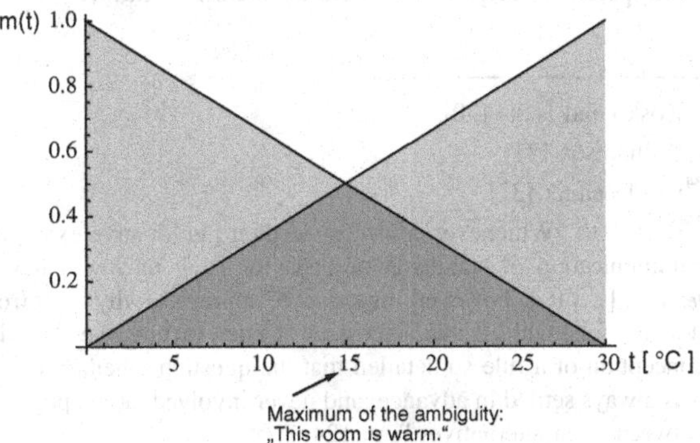

define goals as broadly as possible; for example, a command to the heating control system "Keep this room warm!" cannot be carried out properly when the definition of "warm" is limited to the range 19.5–19.8 °C. The thermostat would continually turn the heating system on and off.

Fuzzy logic is used for example in the subway system in Sendai, Japan; it has been in operation since 1988 without human drivers.[32] Human drivers or conventional computer-controlled systems accelerate or brake the train by observing marks which tell them how far the train is from the station platform. With this kind of control according to a fixed plan, the train's velocity is often abruptly increased or decreased. The Hitachi engineers applied flexible controls in Sendai by making use of fuzzy logic, and these keep the train's motion more uniform. These control rules take into account how often and to what extent the velocity has been changed and how near it is to the maximum allowed velocity. This system has reduced the travel time and energy consumption by 10 % in comparison to a conventional system.

Those philosophers who may be sitting in the train will still not be convinced that an extension of Aristotelian logic to fuzzy logic is acceptable. Thus, T. Williamson has stated that, "Many-valued logic is particularly popular among those who lack logical sophistication".[33] I think that this objection is appropriate if one interprets the membership or opinion functions as *truth functions*.[34] If we take the statement A to be "Max is short" and attribute to it a "truth value" $t(A) = 0.5$, then the statement "Max is short and Max is not short" has a truth value of $t(A \text{ and } \neg A) = 0.5$. This is clearly not what one expects; for the latter case, $t = 0$ would be a reasonable value, since both assertions cannot be true. Similar contradictions are found for the logical conjunction "or". In the weaker form of fuzzy logic using opinion functions, these contradictions do not occur, since opinion functions are not truth functions. I find that half-truths and characterizations such as "very, very true" in contrast to "very true" are not very convincing.

Semantic aspects of indefiniteness play an important role in translating from one language to another and in interpreting legal texts. Vagueness prevents over-regulation and the associated blocking of the system. A reasonable level of control is possible only when a certain tolerance remains to allow for flexible responses. Herbert Lionel Adolphus Hart et al.[35] advocate formulating laws in such a way that they retain an "open texture", and in their interpretation, an adaptive case law can be applied as required by the situation at hand.

---

[32]Kosko and Isaka [30].

[33]Williamson [31].

[34]van Deemter [32].

[35]Hart [33]: "Whichever device, precedent [authorative example] or legislation, is chosen for the communication of standards of behavior [...]; they will have what has been termed an *open texture*. [...] It is, however, important to appreciate why apart from this dependence on language as it actually is, with its characteristics of open texture, we should not cherish, even as an ideal, the conception of a rule so detailed that the question whether it is applied or not to a particular case was always settled in advance, and never involved, at the point of actual application a fresh choice between open alternatives.", p. 124.

On the other hand, power can more easily be (mis)used as a result of vague laws and rules. This is the topic of Michel Foucault's[36] "Deployments of Power". These 'deployments' ('*dispositifs*' in the original French) refer to a mixture of scientific records, administrative directives and regulatory decisions, which are not formulated clearly and are therefore suitable as a means of manipulating people. Foucault himself is not very precise in speaking of them: "[…] un ensemble résolument hétérogène, comportant des discours, des institutions, […], des énoncés scientifiques, des propositions philosophiques, bref: du dit, aussi bien que du nondit, […]".[37] He emphasizes the not-said, which characterizes the indefinite, that which is simply presumed or to be presumed.

How is one to proceed when defining precise limits has serious consequences and there is a need for prompt action, for example in the case of a medical diagnosis? The purely epistomological approach to this problem is based upon the realization that a limit exists, but we do not know where it lies. This attitude represents an optimistic approach to the problem.

It remains unclear whether or not, as a consequence of semantic indefiniteness, the classical theory of concepts itself is in need of revision. This classical theory propounds that concepts are to be defined through sufficient and necessary conditions. Fine-tuning a theory (*tinkering*, *bricolage*) and the tentative nature of theories are strong arguments against a priori or analytic judgments as an element of how we formulate scientific concepts. Claude Lévi-Strauss, in his book "The Savage Mind",[38] describes the *tinkerer*, who can make the best of what is at his disposal and knows how to 'fix things up'. Lévi-Strauss assumes a critical stance which does not sharply distinguish between rational thinking and the primitive (i.e. 'savage') mind.

At the beginning of this chapter, I dismissed the philosophical discussion of vagueness as impractical. That judgment must now be qualified. The artificiality of the scholarly and scientific world corresponds to the limits of our possibilities and abilities to formulate theories, to interpret the world and to change it. True progress demands new concepts and methods. Thus, the progress of the natural sciences is closely linked to the logical-conceptual. Max Bense[39] comments on this connection between logical thinking and practical action as follows: "Now, a statement is the representation of a thought. An application is an action based on a thought. The attention of a logician is focused on the statement, while that of a technician turns to the application. […] Regarding the formal character of a statement, its truth is

---

[36]Foucault [34].

[37]Foucault, *Dits et Ecrits, ibid.*

[38]Lévi-Strauss [35]. "The 'bricoleur' is adept at performing a large number of diverse tasks; but, unlike the engineer, he does not subordinate each of them to the availability of raw materials and tools conceived and procured for the purpose of the project. His universe of instruments is closed and the rules of his game are always to make do with 'whatever is at hand', […]. […] He [the physicist] is no more able than the 'bricoleur' to do whatever he wishes when he is presented with a given task. He too has to begin by making a catalogue of a previously determined set consisting of theoretical and practical knowledge, of technical means, which restrict the possible solutions." pp. 17, 19.

[39]Bense [36].

equivalent to freedom from contradictions; regarding its instrumental character, truth means essentially applicability."

He emphasizes the analogous character of logical and practical actions. To be consistent, one should not underestimate the significance of semantic indefiniteness for the "functioning" or the "failure" of our technological world. A closer examination of the double-slit experiment in physics however also points up the limits of purely logical certainty. Logical necessity is an entirely intellectual construction and cannot replace experience.

Indefinite, in the sense of *vague.*

Examples

- Vague statements and their paradoxical results;
- Fuzzy sets and opinion functions;
- The indefiniteness of opinion functions can be calculated.

2.5 Indistinct Memories of the Past

Although recollections can be stored for many years in our long-term memory, occasionally it happens that invoking them calls up an image which is pale or unfocused in comparison to the original scene. Where and when does indefiniteness enter into memory? Is one impression not uniquely linked to another? Or was a perception simply 'filed in the wrong drawer'? Is the stored information erroneous or vague?

Reconstruction of the past is an attempt to establish causal relations between past and present events. Our memory is the location where the definite and the indefinite from the past are mixed together. We speak of the effort of remembering in order to express the fact that searching our memories for the past is distinct from the subconscious processes which occur in the brain parallel to our conscious thought processes. Therefore, I will include the brain in these considerations of our ability to remember. Neurophysiological research has found that different parts of the brain are active when it is recalling a particular event. The exchange of memory content between different areas of the brain is not understood, and the question remains open as to whether it is controlled by a program. Often, we can only dimly sense a connection between different recollections. One then speaks of implicit memories, which are formed indistinctly and without our conscious intervention, by association or habit. Modern psychology has developed test procedures to analyze impulsive judgments. Imaging techniques make it possible to localize this memory content.

Sometimes we find that we cannot connect a name with a particular face, nor can we remember the place where we saw it for the first time. Psychological experiments have shown that test subjects can recall only about one-fourth of what they had originally committed to memory from a particular experience. They tend to make up new 'information' to complement what is indistinct in their memories. For example, witnesses at a trial may identify the wrong person if he or she is simply wearing the same clothing as the actual person in the past.

Why should we investigate the biophysics of memory before studying its philosophical implications? Our own personal experience can lead us astray when we are dealing with our brains. Wolf Singer[40] has expressed the opinion that we are trained through observations of our surroundings to understand *linear* phenomena, in which a small change in an action gives rise to a corresponding small change in the reaction. In his understanding, we are unable to grasp the complex nonlinear interconnections in the human brain. In trying to comprehend memory processes in the brain, we have to consider a network of 10,000 million nerve cells. Each individual nerve cell (neuron) communicates with roughly ten thousand others which send it input data, and it passes these data on to an equal number of other neurons. This takes place through many interconnections, similarly to the World Wide Web. The shape of the neurons is asymmetric. So-called dendrites form finely branched input connections. A long axon conducts the impulses as excitation potentials in the range of millivolts along the length of the nerve cell. Insofar as an axon of cell A repeatedly excites another cell B, the cells modify themselves so that the effectiveness of the excitation is amplified. The strength of the neural interconnections is thus fundamental in the storage of recollections. Learning is based on a fine adjustment of the connections (synapses) between the nerve cells, which transfer the excitations from cell to cell. They permit the recognition of patterns that have been stored in the network. Descartes' "*res cogitans*" is not a computer. The brain functions in a completely different fashion from a digital computer in which every bit of information is allocated to a particular storage site.

Neurons fire at intervals of milliseconds, i.e. they emit short signals (action potentials) which lead to transport processes in the ion channels of other neurons which are interconnected to them. Positive ions flow through the interior of the nerve cells and increase the potential there. The basic equation of associative memory considers the rate of firing of each neuron in the network as a function of the potential of that neuron. This equation becomes a closed system when one takes into account that the potential of each neuron itself depends linearly on the rate of firing of other neurons that are providing input to it. Learning means adjusting the strengths of the synapses. Positive and negative synaptical strengths correspond to excitation or damping of neuronal stimulation.

Neurophysiological research attempts to understand how the still-undefined individual content from the various sensory organs is combined into a recognizable pattern or sensory object. Spatial, color or nonvisual—i.e. heard or felt—impressions arrive in different areas of the cerebral cortex and are assembled together in a process which involves the frontal lobes (basal ganglia) and the thalamus (near the brain stem). This synchronizes the various nerve signals. In the case of visual

[40]Singer [37].

information, the synaptic connections take the form of maps, whose topology corresponds to the correlations of the input signals. Gerald Maurice Edelmann[41] has proposed the hypothesis that the reciprocal exchange between such maps provides support for the synchronization of memories. He has developed the concept of *reentry* which he defines as the recursive interchange of signals that occurs in parallel between brain maps, and which continuously interrelates these maps to each other in time and space.

In contrast to the philosophy of the mind, neurophysiology sees consciousness as an indefinite correlate of a dynamic neuronal core structure, which is defined through causal relations. The variability of this core structure permits the formation of "attractors", i.e. patterns towards which the brain gravitates almost independently of the particular initial conditions. Memory content can be considered to comprise such patterns. By creating such attractor patterns, our brains can adapt to the requirements of a permanently changing environment.[42] The freedom of humans thus resides not in the possibility of our "freely" making decisions; it is rather to be found in the individual structuring of our likes and dislikes. Does this creativity also allow the memory to combine indefinite bits of content into scenarios which produce a coherent past? Indeterminacy appears here in individual reconstructible processes whose combined action is beyond our understanding. Wolf Singer asserts that "The dynamic states of billions of neurons in the cerebral cortex display a degree of complexity which is far beyond anything that we can imagine. That does not necessarily mean that we will not be able or are not willing to develop analytical methods which can identify these systemic states and follow their temporal evolution; their description will however be abstract and vague and will bear no similarity to our everyday ideas and concepts, which themselves are based upon the functioning of these neuronal states." Interdisciplinary cooperations can focus on the intersection "consciousness/preconsciousness". Scholars of the humanities and the social sciences investigate and interpret consciousness and its cultural concretions, while neurophysiologists study the preconscious.

At the beginning of the past century, various psychologists including Sigmund Freud, Carl Gustav Jung, Alfred Adler and others were attempting to bring structure into the indefinite aspects of our memory. They started with empirical case histories which they had encountered in their clinical practice as psychotherapists. Freud assumed that a part of our soul, the indefinite Es, lies hidden from the conscious Ego. He called this part the *unconscious* and made it into a fundamental component of psychoanalysis. The Freudian unconscious is accessible to the psychoanalyst only when he can bring it into the conscious mind. The psychiatrist and philosopher Thomas Fuchs[43] commented on the process of internalization from the neurobiological viewpoint: "Every contact with others leaves traces on the neuronal level through synaptic learning; to be sure, not necessarily in the form of localisable

---

[41]Edelman [38].

[42]Edelman: *Wider than the Sky*, *ibid.*, pp. 145 ff.

[43]Fuchs [39].

"memories", "images", or "representations" of the interactions or the contact persons, stored in fixed locations, but rather as dispositions or perceptions, feeling and behavior." In the collective consciousness of groups, individual experiences are generalized and stored with the help of symbols. History deals with such memories and their concretization in historical documents. The options of historians are however limited by the brief horizon of memory. "The true image of the past scurries away rapidly; we can capture it only as a picture which flashes up fleetingly at the moment of its perceptibility and then disappears forever. [...] For it is an unrecoverable image of the past that threatens to vanish in every present which fails to recognize itself as reflected in that past."[44]

Closely related to remembering is forgetting. While memory can be investigated in experiments with animals, it is more difficult to study forgetting in that way. Forgetting is not only a weakening of recollections, but is often an active suppression of unpleasant, unwanted memories. The systems analyst Niklas Luhmann considers forgetting to be a kind of indefiniteness which is produced by our memory itself, as a creative act which opens up space for new possibilities of creative choice. Here, we can discern similarities to the collective forgetting by a whole society, which is called into play through disagreeable or painful events. It is however questionable whether such forgetting can lead to a genuine renewal.

On the basis of recent results from brain research, the historian becomes a kind of knowledge archeologist, who is confronted with contradictory recollections. The historian Johannes Fried[45] describes in his book "*Schleier der Erinnerung*" how the two physicists Werner Karl Heisenberg and Niels Bohr retained quite different memories of their meeting in Copenhagen in 1941. From their reports, Fried finds, the meeting took place "at an undetermined time, in an undetermined place, under undetermined circumstances, and dealt with undetermined topics." We are accustomed to the fact that the historical fragments from ancient times often contain little original knowledge, but stand at the origins of well-thought-out stories and legends. History, which deals with memories and recollections, has to accept[46] that these become less and less clear with the passage of time, i.e. they become indistinct. The objects of history are unique; their reconstruction entails indefiniteness.

Indefinite, in the sense of *indistinct*.

Examples

- Calling up memories and recollections;
- Synapses learn dispositions;
- The reconstruction of the historical past.

---

[44]Benjamin [40].

[45]Fried [41].

[46]Schneidmüller [42].

2.6 Indefinite Ontology

What is the origin of the universe? Is there a reason why the world came into being? I call such questions and their respective indefinite answers *ontological*, since they are related to our image of the whole of the world (all of existence, "being"); they are not limited to just parts of it. Are these questions well defined? As scientists, we are accustomed to treating such questions with caution. In this chapter, I will carry out the "experiment" of comparing the modern physical theory of *cosmogeny*, the formation of the world, with the philosophical thinking of Plato on the foundations of the world. Ancient Greek philosophy has always been of great interest to physicists, since many of our modern concepts, such as that of the atom, first arose at that time and place. I have chosen the ancient philosopher Plato (424-348 B.C.) because the indefinite and the definite played a central role in his thought. The comparison will also illustrate just how differently physics and philosophy go about searching for the origins of the universe.

Let us follow the history of our universe, presumed to be (13.5–14) billion (10^9) years old, to its beginnings, by taking a look backwards; we choose the moment of the Big Bang as the zero point of our time scale. Then we find the birth of our galaxy around a billion years later. Up to 300,000 years following the origin of the universe, it is still opaque; i.e. there are no electromagnetic signals from this early era. At a temperature of 3000 K, a thousand times hotter than the current temperature of interstellar space, electrically neutral atoms can form by the combining of electrons and protons (0 °C corresponds to 273 K, room temperature to 293 K). Light was then free to move outwards to the boundaries of the universe, since its energy was not high enough to ionize the atoms and it was therefore no longer absorbed by them. This is termed the *decoupling* of radiation and matter. At the time the atoms were formed, the light in the universe had a Planck spectrum similar to that of the sun today (corresponding to 5800 K), just somewhat cooler. The earlier we examine the universe, the hotter it was; i.e. the greater the energy scale which plays a role in the processes taking place. The atomic nuclei were formed one second after the Big Bang, at a temperature of 10^{10} K. The neutrinos play an important role in the synthesis of nuclei, and the existence of three light neutrinos in the Standard Model agrees well with the distribution of the elements throughout space. In the early universe, the macroscopic astrophysics of the universe and the microscopic physics of elementary particles act together, since the increasingly high temperatures permit higher and higher energetic excitations. Some microseconds (10^{-6} s) after the Big Bang, the quarks and gluons combined to form nucleons. Going back to 10^{-10} s, we can apply knowledge obtained and tested in laboratory experiments in order to describe the hot fireball of the early universe, which had a temperature at that time corresponding to 100 GeV (10^{15} K). This corresponds to our state of knowledge including the recent discovery of the Higgs boson. One expects that at still higher temperatures, all the known interactions would have been equally strong. Quantum field theories contain virtual processes which renormalize the couplings. Through these corrections, the strength of the couplings varies with resolution or energy, leading to an increase of the weak coupling constants and a decrease of the strong coupling at high energies or temperatures.

The hypothesis of an initial point at which time and space were created, which is in common use by cosmologists today, is controversial in terms of the question of just how singular that beginning really was—was there already a 'time' at the moment of the Big Bang? The higher the energy density of the universe, the stronger the curvature of space and time. At the moment of the Big Bang itself, one can therefore no longer speak of space and time. The quantum effects on gravitation which would then occur are currently not understood.

The cosmic background radiation which was produced at the time when atoms formed can be observed today at a temperature of slightly below 3 K. This low temperature arises from the thousand-fold expansion of space since the time when this radiation appeared. The temperature of the background radiation is constant down to deviations of one part in 100,000, and it thus gives precise information about the density fluctuations within the universe, which gave rise to the galaxies. The measurements indicate a flat universe, with no curvature. How can a flat universe correspond to a solution of the Einstein equations, although that solution is particularly unstable and requires a very carefully chosen initial value for the energy density of the universe? Why is the background radiation so homogeneous? Various spatially separated parts of the universe could hardly have been in close (causal) contact at the moment of decoupling after the Big Bang. Why should they now have the same temperature? The explanation requires a mechanism which is still not fully understood by present-day theory. The Einstein equations permit an exponentially increasing expansion of the universe, in the case that there is a scalar field which evolves with an extremely small kinetic energy, so that its predominantly potential energy can drive the expansion; the latter is then extremely fast. Within 10^{-35} s, the universe would expand by many orders of magnitude. The scalar inflationary field blows up the universe so rapidly that parts of the universe which are widely separated by the time of the formation of atoms could still have been in causal contact with each other. This would explain the extreme homogeneity of the 3 K-microwave radiation.

Cosmologists have estimated from the mass distributions of the spiral galaxies that the energy density of the visible universe is only 4.8 % of the critical mass density required for a flat universe. There must therefore be non-observable matter, known as *dark matter*. It is called dark because it neither reflects nor emits nor absorbs light. From the combined analysis of galactic clusters, supernovae and the microwave background radiation, an upper limit for the percentage of dark matter which contributes to the critical density can be found. The result is that this cold, dark matter contributes 26.2 % to the critical density. The rest (69 %) is attributed to an either weakly time-dependent or time-independent *dark energy*. Albert Einstein introduced a cosmological constant into his original equations of general relativity (to make the universe stable, since its expansion had not yet been observed at the time); this constant can accurately play the role of the dark energy. The value of the cosmological constant can be determined from observations of the accelerating expansion of the universe. Its magnitude and sign are not yet theoretically understood. We have already pointed out this problem in discussing the results of string theory (Sect. 2.1). The increase of our knowledge has led to an increase of the boundaries of that knowledge. The dark, unobserved parts of the universe now constitute 95.2 % of the whole. These results were obtained by satellite measurements from the cosmic microwave spectrum in combination with lensing reconstruction and other external data.[47] Dark matter bends the light rays of galaxies and makes multiple images of the same background source appear. This effect, known as "gravitational lensing", can be used to map out the distribution of dark matter.

Great indefiniteness calls up a need for more knowledge and less uncertainty, and it thus spurs on efforts to reduce the indefiniteness. Precisely this effort to escape from the fog of indefiniteness was the driving force for Plato in his late work "*Philebus*", in which he declares the definite and the indefinite to be fundamental components of the universe. One could find here an analogy to modern physics, which presumes 4.8 % definite and 95.2 % indefinite matter/energy in the universe at present. But let us look deeper into Plato's philosophy.

The following figures play roles in "Philebus": *Socrates*, who asks questions and instructs; the knowledgeable and flexible *Protarchus*; and, in the introduction, *Philebus*. Socrates begins the trialog by citing the thesis of the Hedonist Philebus, to the effect that well-being, pleasure and enjoyment are good for all living beings. He counters this thesis by maintaining that wisdom, correct opinions and true reasoning are more important. The conversation seems to revolve around the proper mixture of wisdom and pleasure. Then suddenly the main theme emerges—there are four original components of being, four elements: the indeterminate (*apeiron*), the determinate (*peras*), a mixture of the two (*meixis*), and the cause of the union of the indeterminate and the determinate (*aition*). The indeterminate can vary gradually and is thus unlimited. "Everything of which we see that it becomes more or less, and strong or weak, and all similar properties, all that must be collected under the category of the unlimited (the indeterminate) in unity." Today, we would speak of *continuous* properties.

That which is defined by a limit, in contrast, is clearly distinguished from its opposite; it has *discrete* properties: "Thus that which does not take on those properties, but rather everything opposite to them, starting with the equal and equality, and after equality twice as much and every whole number in a ratio to numbers, and every measure in a ratio to measures, if we count all that among the limited (the determinate), we would be on the right path".[48]

Plato in "Philebus" explains the indeterminate, which we have called the indefinite, undefined or unknown (cf. Sect. 1.2). He sets the finite and defined as an opposite to the indefinite, as a class of concepts or objects which can be clearly identified and enumerated with numbers. The determinate belongs to a class (paragraph 25c) which puts an end to all differences and opposites.

In the class of the determinate objects, there is no discussion as to whether a particular element belongs to it or not, but rather a clear and unique assignment can

---

[47]Planck satellite: 2015 results. XIII. Cosmological parameters. Table 4, *astropreprints* 1502.01589.

[48]Plato, *Philebos*, 25b. All the quoted paragraphs correspond to those in the German edition, Hamburg (2007), pp. 442 ff. For the Greek and English versions, see e.g. *Philebus*, http://www.perseus.tufts.edu/hopper/text?doc=Perseus%3Atext%3A1999.01.0174%3Atext%3DPhileb.%3Asection%3D25b. The paragraph numbering is similar.

be made of a first, second and third element of this class (25c). Here we find a veneration of numbers which recalls the presocratic Pythagoras (570-495 B.C.), where numbers determine among other things the fundamental intervals in music. The octave, the fifth and the fourth are defined by the ratios of numbers (2:1), (3:2) and (4:3) on a vibrating string. If one were to change the length of the string continuously, corresponding more or less to the indefinite, there would be no harmonic overtones. "And in the acute and the grave, the quick and the slow, which are unlimited, the addition of these same elements (the determinate, discrete, finite) creates a limit and establishes the whole art of music in all its perfection, does it not?" (26a).

Protarchus develops the discussion further by recognizing that a mixture of the indeterminate with the determinate represents a new formative element: "Thou evidently wish to say that if these two are mixed, certain results are produced in each instance".[49] Socrates agrees: "And thence arise the seasons and all the beauties of our world, when the unlimited and that which is limited within itself are mixed together" (26a). Plato wants to understand the mixture as a new unit or element; he sees the process "from Becoming to Being" (26d).

Then Socrates begins to speak of the *cause*: "But we said there was, in addition to three classes, a fourth to be investigated. Let us do that together. See whether you think that everything which comes into being must necessarily come into being through a cause" (26e). Physics recognizes causes, which always precede their effects.

Students of physics encounter *causes* in Newton's first law of motion, which maintains that a body which is not acted upon by any external forces remains at rest or in a state of uniform linear motion. Later, when they attempt to derive electromagnetic radiation (from Maxwell's equations), they encounter the principle of causality. This principle states that effects can propagate (from their causes) with at most the velocity of light. Thus in spacetime, only events within the so-called past light cone can influence events in the present; likewise, the absolute future is limited by the future light cone. Since there are very many equivalent inertial frames of reference, special relativity theory states that an event can never be associated with a particular time coordinate. But cause and effect maintain the same temporal order under Lorentz transformations, which conserve the structure of spacetime.

The "Platonic cause" is to be distinguished from physical causes; it can to be sure also play that role within causality, but it has a still deeper significance. 'Nothing is without a cause' could be a brief summary of this principle. Here, the final cause ("*causa finalis*") is meant, as it represents the purpose of the universe. In Plato's work "*Timaeus*", Timaeus explains the formation of the world in all its details. At the beginning, he warns: "Wherefore, Socrates, if in our treatment of a great host of matters regarding the Gods and the generation of the Universe we prove unable to give accounts that are always in all respects self-consistent and

[49] Plato, *Philebus*, *ibid.*, 25e.

perfectly exact, be not thou surprised".[50] Read naively, fire, earth, water and air become the original components or elements—in complete agreement with traditional ancient cosmology. But the real elements are the forms, namely triangles (53d). In Timaeus, we recognize the theme of Philebus once again: "Midway between the Being which is indivisible and remains always the same and the Being which is transient and divisible in bodies, He (the Demiurge) blended a third form of Being compounded out of the twain" (35a).

Paul Natorp[51] interprets this cosmogeny in terms of the paired but opposite concepts represented by the indeterminate and the determinate. He understands their combination as an 'ensoulment' or a harmonization. In his book "*Platos Ideenlehre*", he explains the fourth element, the *cause*, or *reason*, as an additional ingredient to complement the other three. "Cause is in general none other than the law or the determinant of the indefinite". Natorp explicates the aspects of the Philebus text in poetic language.

Plato's emphasis of the quantitatively indefinite fascinates me, along with its opposite, the mathematically-ordered definiteness. Here, I see a line of development leading to structuralism, which sees symbols as a mixture of significant data, an indefinite object, and theory. In the case of physical and biological concepts and theories, one can readily comprehend[52] how new knowledge improves our definitions of objects. How, though, does the determining element mix with the indeterminate? Is there a Platonic cause for this?

What is the Platonic cause? In this chapter, I have attempted to show that one has to count on indefiniteness, but that one can also count with it. An essential point was the distinction between knowledge (e.g. astronomy) and "interpretation" (e.g. astrology). If we can explain only 4.8 % of the energy density of the cosmos with the Standard Model of elementary particles, then a large amount of the indefinite remains, and it cries out for continued research. These indefinite elements are necessary in order to arrive at a consistent description of the observed flat universe through Einstein's equations. The mixture of definite and as yet indefinite physics is to be found in gravitational theory. The astronomer-physicist is not searching in vain.

The philosopher may want to challenge the cause, or reason: "Is the final word that we can say of being: 'Being means reason'? Or does not the essence of humanity, does not our belonging to being, does not the nature of being yet remain, and they remain still the disconcertingly memorable? Dare we, if it should be so, yield up this most memorable in favor of simply calculating thought and its enormous successes? Or is it not our responsibility to find paths on which our thinking can answer to the memorable, instead of sneaking past it, bewitched by calculating thought? That is the question. It is the universal question of thinking. Its

---

[50]Plato: *Timaios*, Hamburg 2007, 29c. English: e.g. *Timaeus*, http://www.perseus.tufts.edu/hopper/text?doc=Perseus%3atext%3a1999.01.0180%3atext%3dTim.

[51]Natorp [43].

[52]Pirner [44].

answer will decide what is to become of the Earth and of the existence of humanity on this Earth".[53]

This quote from Martin Heidegger would seem to engender both emptiness and a deep meaning at the same time. Have the determinate and the indeterminate mixed together in his thinking to yield something genuinely new? Or do the definite thoughts which he produced just point to a certain origin and thereby make their own value questionable? I for one cannot make head nor tail of this philosophy, because its association to Being is circumscribed only very vaguely. A more balanced view of Heidegger's involvement with the natural sciences and natural philosophy is given in an article by Joseph Rouse, who states that recent examples in the history of physics "cannot be appropriately regarded as impositions of a predetermined orientation toward calculative control upon nature as a plastic resource, for what it matters to understand collectively and what is at stake has shifted. Such shifts instead reflect an openness within science to allowing things to show themselves intelligibly in new ways".[54] Modern cosmology is yet another example.

Indefinite, in the sense of *undefined*.

Examples

- Dark matter and dark energy in cosmology are (still) undefined;
- The Undefined, the Unlimited in Plato's philosophy;
- The mixture of the indefinite and the definite.

References

1. Bohigas, O., Weidenmüller, H.A.: Aspects of chaos in nuclear physics. Ann. Rev. Nucl. Part. Sci. **38**, 421–453 (1988)
2. Šeba, P.: Parking in the city. Acta Physica Polonica A **112**(4), 681 (2007)
3. Abul-Magd, A.Y.: Modeling gap-size distribution of parked cars using random-matrix theory. Physica A **368**, 536 (2006)
4. Nielsen, H.B., Brene, N.: Some Remarks on Random Dynamics. http://www.nbi.dk/~kleppe/random/Library/remarksonrd.pdf
5. Ellis, G.F.R., Smolin, L.: The Weak Anthropic Principle and the Landscape of String Theory (2009). arXiv:0901.2414v1[hep-th]
6. Bousso, R.: Pontifical academy of sciences. Scripta Varia **119** (2014) (Vatican City)

[53]Heidegger [45].

[54]Rouse [46].

7. Bousso, R., Polchinski, J.: Scientific American, p. 79 (2004)
8. Duhem, P.M.M.: La théorie physique: son objet, sa structure, Paris, 4eme édition, p. 331 (1997) (own translation)
9. Lorenz, E.N.: Climate Predictability. The Phys. Basis of Climate and Climate Modelling **16**, 132–136 (1975) (WMO GARP Publication Series)
10. Hasselmann, K.: Is climate predictable. In: Bunde, A., Kropp, J., Schellnhuber, H.J. (eds.) The Science of Disasters—Climate Disruptions, Heart Attacks, and Market Crashes, pp. 141 ff. Heidelberg (2002)
11. Pareto, V.: Cours d'Économique Politique, vol. 2. London (1897)
12. Levy, M., Solomon, S.: New evidence for the power-law distribution of wealth. Physica A **242**, 90–94 (1997)
13. Clementi, F., Gallegati, M.: Power Law Tails in the Italian Personal Income Distribution (2004). arXiv:cond-mat/0408067v1[cond-mat.other]
14. Mantegna, R.N., Stanley, H.E.: Investigation of Financial Markets Using Statistical Physics Methods. In: Bunde, A., Kropp, J., Schellnhuber, H.J. (eds.) The Science of Disasters—Climate Disruptions, Heart Attacks, and Market Crashes. Heidelberg (2002)
15. Sinha, S., Chatterjee, A., Chakraborti, A., Chakrabarti, B.K.: Econophysics, Weinheim (2011)
16. Bouchaud, J.P.: The (unfortunate) complexity of the economy (2009). arXiv:0904.0805v1 [q-fin.GN]
17. Mittelstaedt, P.: Philosophische Probleme der modernen Physik, p. 146. Munich (1963)
18. Hornberger, K. et al.: Collisional decoherence observed in matter wave interferometry. Phys. Rev. Lett. **90**, 160401 (2003). arXiv:quant-ph/0303093v1
19. Joos, E., Zeh, H.D.: The emergence of classical properties through the interaction with the environment. Zeitschrift für Physik B **59**, 223 (1985)
20. Paz, J.P., Zurek, W.H.: Environment-Induced Decoherence and the Transition From Quantum to Classical, Lectures given by both authors at the 72nd Les Houches Summer School on "Coherent Matter Waves", July–August 1999. arXiv:quant-ph/0010011v1
21. Bohm, D., Hiley, B.J.: The undivided universe: an ontological interpretation of quantum theory, p. 31 ff. Routledge, London (1993)
22. Padmanabhan, T.: Gravitation: Foundations and Frontiers, pp. 670 ff. Cambridge (2010)
23. Susskind, L., Lindesay, J.: An Introduction to Black Holes, Information and the String Theory Revolution: the Holographic Universe. Singapore (2005)
24. Kemmerling, A.: Informationsimmune Unbestimmtheit, Bemerkungen und Abschweifungen zu einer offenen Wunde der theoretischen Philosophie. Lecture given at the Marsilius Kolleg (unpublished)
25. Bak, P., Tang, C., Wiesenfeld, K.: Self-organized criticality: an explanation of $1/f$ noise. Phys. Rev. Lett. **59**, 381–384 (1987)
26. Frette, V., Christensen, K., Malthe-Sørenssen, A., Feder, J., Jøssang, T., Meakin, P.: Avalanche dynamics in a pile of rice. Nature **379**, 52 (1996)
27. Zadeh, L.A.: Fuzzy Sets. Inf. Control **8**, 338–353 (1965)
28. DeLuca, A., Termini, S.: A Definition of a Nonprobabilistic Entropy in the Setting of Fuzzy Sets Theory. Inf. Control **20**, 301–312 (1972)
29. Bandemer, H., Näther, W. Fuzzy Data Analysis, p. 46. Dordrecht (1992)
30. Kosko, B., Isaka, S.: Fuzzy logic. In: Scientific American, p. 76 (1993)
31. Williamson, T.: Précis of vagueness. Philos. Phenomenol. Res. **LVII**(4), 922 (1997)
32. van Deemter, K.: Not Exactly: In Praise of Vagueness, p. 193 ff. Oxford University Press, Oxford (2010)
33. Hart, H.L.A.: In: Bulloch, P.A., Raz, J. (eds.) The Concept of Law. Oxford (1997)
34. Foucault, M.: Dits et Ecrits, III, p. 299, (1977)
35. Lévi-Strauss, C.: The Savage Mind. University of Chicago Press, Chicago. (Title of the original edition: *La Pensee Sauvage*, Paris 1962) (1966)
36. Bense, M.: Die Philosophie, Zwischen den beiden Kriegen. Frankfurt am Main, p. 62 (1951)

37. Singer, W.: Understanding the brain. How can our intuition fail so fundamentally when it comes to studying the organ to which it owes its existence?, EMBO reports 8, S1, S. 16–19 (2007)
38. Edelman, G.M.: Wider than the Sky: The Phenomenal Gift of Consciousness. Yale University Press, New Haven (2004)
39. Fuchs, T.: Das Gehirn—ein Beziehungsorgan. Eine phänomenologisch-ökologische Konzeption. Stuttgart, p. 190, (2009)
40. Benjamin, W.: *Thesen zur Philosophie der Geschichte, in: Illuminationen* (V), (Frankfurt/Main. English translation by Zohn, H. Illuminations: Essays and Reflections) ed. Hannah Arendt (1969)
41. Fried, J.: Der Schleier der Erinnerung. Grundzüge einer historischen Memorik. Munich, p. 31 (2004)
42. Schneidmüller, B.: Die Akzeptanz von Unbestimmten als Prinzip der historischen Methode (lecture at the Marsilius Kolleg, unpublished)
43. Natorp, P.: *Platons Ideenlehre*, first edition 1903, new edition Hamburg 2004, p. 331 (Natorp, P.: Plato's Theory of Ideas. ed. Politis, V.; transl. Connolly, J.). Akademia Verlag, Sankt Augustin (2004)
44. Pirner, H.J.: The semiotics of postmodern physics In: Ferrrari, M., Stamatescu, I.O. (eds.) Symbol and Physical Knowledge. On the Conceptual Structure of Physics, Heidelberg, pp. 211–227 (2001)
45. Heidegger, M.: *Der Satz vom Grund*, Heidegger Complete edition, vol. 10, Petra Jaeger, Frankfurt am Main, 1997, p. 189. (English: Heidegger: The Principle of Reason, Indiana University Press, 1996)
46. Rouse, J.: Heidegger on Science and Naturalism. Division I Faculty Publications, Paper 36 (2005). http://wesscholar.wesleyan.edu/div1facpubs/36

Chapter 3
Approaching the Definite

The acquisition of information can be considered as an *approach* to the definite. The investigation of the various manifestations of indefiniteness which we presented in Chap. 2 showed that they can be divided into two groups. The elements of the first group describe our *lack of knowledge*, i.e. they appear arbitrary, uncertain and indeterminate. The elements of the second group demonstrate our *inability to comprehend*; they are vague, blurred and undefined. In order to eliminate the indefiniteness of the first group, we need more knowledge, more information. The present chapter thus takes up the theory of information. When is the average information content of two signs maximal? How can a not-yet-determined text be classified with the help of information theory? I will discuss how a "system" can react in a flexible manner to the indefiniteness of its environment, by extracting information from its surroundings and thus increasing its internal complexity. Ecology offers many examples of this, e.g. a pond full of organisms whose life histories are determined by the sun, the air and possibly by inflows to the pond. It is important to pin down the flow of energy and materials. Energy once used by the ecosystem is converted into heat, and in this degraded form can no longer sustain life processes. The one-way flow of energy and the circulation of materials are two principles of general ecology.

We will also encounter the controversial idea of *complexity*. This chapter introduces as a new concept the "value" (worth, quality) of information and makes an attempt to establish the dynamics whereby information, complexity and indefiniteness mutually interact. We will be guided by the history of thermodynamics, which attributes a value or quality to energy, depending on how effectively it can be converted into useful work. In the *information age*, we are concerned with the value or quality of *information*. Such a judgment of the quality of information is possible because the definite system and its indefinite environment are considered together. Thus, this chapter deals with a variation on the principal theme of this essay: considering the definite and the indefinite together. Through information, one gains a *definite* impression.

H.J. Pirner, *The Unknown as an Engine for Science*,
The Frontiers Collection, DOI 10.1007/978-3-319-18509-5_3

But information cannot remedy the indefiniteness of the second group; thus the process of acquiring information defines only an *approach* to definiteness. In the following Chap. 4, I will then take up the topic of how to deal with the second group, the vague, blurred and undefined.

3.1 Information

Our knowledge is on the one hand limited by the unknown, on the other however also by the *undetected* and its structures. We cannot know what we do not know. Nevertheless, we can hope that there is a reservoir of potential information into which we can tap. The historian dreams perhaps of yet-undiscovered source documents which are slumbering in the archives. The natural scientist mulls over new experiments which could elucidate relationships that are still only dimly suspected. But it can also happen that the information which we already possess simply needs to be regarded in a new light. When information meets up with the indefinite, they evolve their own dynamic, called *information dynamics*,[1] which we shall now examine more closely. Our starting point for this investigation is the question: What is information? What is determined by information?

The concept of *information* was mathematically defined in 1948 by the mathematician and engineer Claude Elwood Shannon. In his work on the theory of communications, he characterized information in terms of the possible results of transmitting a signal. His concept of information begins with a source which transmits randomly-distributed information. The information which is contained in each symbol is then a function of the frequency of occurrence of that symbol. Let us assume that we transmit a string of six characters, A B B B B A; then A occurs twice and B four times. From these frequencies of occurrence, one can compute the probability of occurrence for each letter, i.e. the probability for the letter A is $p_A = 2/6$, and for B, $p_B = 4/6$. The definition of information is based on two fundamental assumptions:

(i) A symbol with a lower probability has a higher potential *information content*. It is more difficult to guess that the next letter to be transmitted will be A, since the letter B occurs more often. The surprise effect is greater when one receives the letter A. Thus, one ascribes a higher information content to A:

$$I(p_A) > I(p_B) \text{ if and only if } p_A < p_B.$$

(ii) The information from two independently-transmitted communications is additive. Let us consider an example of this principle: Let the probability that John is driving his car be $p_A = 60\ \%$, and the independent probability that he will receive a call on his mobile phone be $p_B = 10\ \%$. Then we can compute

[1] Shannon [1].

the probability that he will receive a call while driving as the product of the two; this combined probability is thus $p_A p_B = 0.06$ or 6 %. When these two completely independent events occur together, the surprise effect is greater than for the occurrence of each event alone, i.e. the information content of the two adds:

$$I(p_A\, p_B) = I(p_A) + I(p_B).$$

These two assumptions (i) and (ii) together show that the information is proportional to the *logarithm* of the inverse (!) of the probability. Since the logarithm is a monotonically increasing function, rare occurrences whose inverse probability ($1/p$) is large have a high information content. Historically, the logarithm was defined in order to transform difficult multiplications into easier addition problems by the use of tables; it therefore fulfils the second basic assumption (ii). If we make use of logarithms to the base 2, the quantity of information thus defined is the *bit*[2]:

$$I(p) = \log_2(1/p) = -\log_2(p).$$

The information which is contained in a yes-no decision with $p(\text{yes}) = p(\text{no}) = ½$ is $I(1/2) = 1$ (bit). The text which is thus far stored in my computer consists of 256 kBytes,[3] i.e. 2,048,000 yes-no decisions (bits). With 32 independent symbols in the alphabet, each symbol has an information content of 5 bits. One can try to guess the next symbol which will come in a text that is known up to a certain point. The information content of a symbol is then found from the average number of guesses required. The average information content of a symbol is 1.5 bits, and is thus less than 5 bits. Therefore, one can compress a text by a factor of roughly 3 without changing its information content.

The *mean value* of the information, $\langle I(p)\rangle$, is obtained as the sum over all possible bits of information with their associated probabilities, which add up to the overall probability of $\sum p_i = 1$:

$$H = \langle I(p)\rangle = \sum p_i I(p_i) = -\sum p_i \log_2(p_i).$$

This quantity H is known in physics and chemistry as *entropy*, and we have met it already in a similar but not identical form in the definition of semantic indefiniteness in Sect. 2.4. (The difference is due to the fact that opinion functions are not the same as probabilities.) Entropy assumes its maximum value in the extreme case of a uniform distribution. A glass of water with an ice cube has a lower entropy than the same glass after the ice cube has melted, i.e. when the same temperature applies everywhere within the glass. We know from thermodynamics that a system with a higher entropy has less free energy, i.e. less energy that can be converted into useful work.

---

[2]In particular, for logarithms to the base 2, we see that $\log_2(2) = 1$, i.e. $\log_2(1/2) = -1$.

[3]1 Byte = 8 bits.

In the mathematical theory of information, a large value of H is associated with a high degree of indefiniteness of the data. A string of characters with the two symbols A and B which occur with the probabilities $p_A = p$ and $p_B = 1 - p$ has an average information content H given by:

$$H = -\mathrm{p}\log_2(p) - (1 - \mathrm{p})\log_2(1 - p).$$

In Fig. 3.1, the average information content of these two symbols is shown as a function of p. If the probability of occurrence of symbol A given by $p_A = 0$, then only the second symbol B can occur and the surprise effect on receiving a character string which consists only of B's is zero. If, on the other hand, the two symbols occur with the same probabilities, $p_A = p_B = ½$, then the surprise effect is maximal when we receive either the symbol A or B.

A pair of dice has a greater number of possible results than a coin; therefore its average information content is higher. If the 6 faces of a die are equivalent, each face i occurs with the probability $p(i) = 1/6$. Summation over the 6 faces then yields $H = 2.58$.

After reading the first 100 pages of a crime novel, I might well ask the question as to whether Mr. Smith is the murderer, or instead Mrs. Murphy. This probability is however differently defined from the probability that we are using here. It depends on my subjective impression. In daily life, such conjectures are often raised, for example in a courtroom. Each member of the jury may interpret the evidence differently and arrive at a different conclusion. It is therefore desirable to have a jury with a number of members to reach a verdict fairly.

The Bayesian theory of probability takes the prior assumptions which enter into estimation of the probabilities into account. This is possible by defining *conditional probabilities* $P(x|y)$, which express the probability that an event x will take place if the event y has already occurred with certainty. This condition can be made clearer by an example: In every language, certain combinations of letters are particularly common, e.g. in English, q is always followed by u, so that $P(\mathrm{u}|\mathrm{q}) = 1$. This conditional probability should be distinguished from the *combined* probability

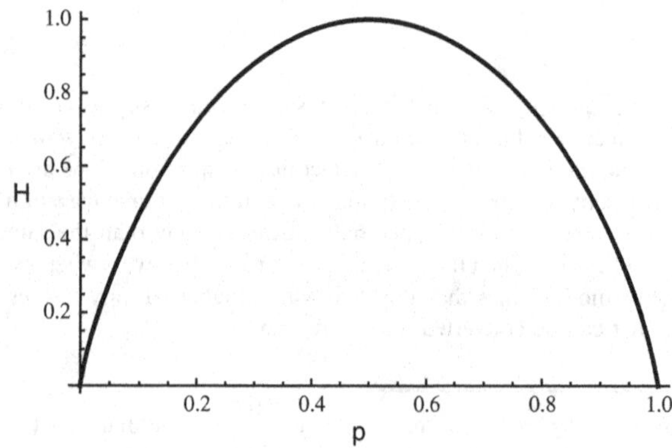

Fig. 3.1 The average information content H in bits for the case of two symbols A and B which are transmitted with the probabilities $p_A = p$ and $p_B = 1 - p$. The surprise effect is maximal when the two events have the same probability ($p_A = p_B = 1/2$)

$P(x, y)$ that *both* x and y occur. Using the Bayesian theory, the conditional probability $P(y|x)$ can be expressed in terms of the probabilities $P(x|y)$ and $P(y)$:

$$P(y|x) = \frac{P(x|y)P(y)}{\sum_{y'} P(x|y')P(y')}$$

Conditional probabilities are useful for example when questions of the following type are to be answered: Karl is to be tested for Spanish influenza. He belongs to a low-risk group who become infected with influenza with a probability of only 1 %. The test has a detection probability of 95 %. The test turns up positive for Karl. What is the probability that Karl in fact has the Spanish flu? A simple calculation[4] shows that this probability is 16 %. The conditional probability of 95 % is reduced by the small risk of his group. This computation can be understood by applying the above formula, or more simply by considering the occurrence frequencies. Without loss of generality, we can assume that the risk group to which Karl belongs consists of 10,000 persons. In this group, 100 persons, i.e. 1 %, will contract the Spanish flu, and 9900, or 99 %, will not. Owing to the imperfect detection probability of the test, 95 % or 95 persons out of the 100 infected will test positive and 5 will show a false negative result. Similarly, for the 9900 non-infected persons, the test results will show false positive results for 5 % (495) persons, and 9405 will be correctly identified as not infected. Now what does the positive test result for Karl mean? The probability that Karl has the Spanish influenza is obtained from the relative frequencies of persons who correctly test positive in relation to the total number of persons from his risk group who test positive, $P = \frac{95}{95+495} \sim 16\%$.

A good approach for reducing uncertainty is to ask for the relationships between different pieces of information. If one has several texts whose authors are unknown, assumptions about their similarities with texts from known authors can be supported or refuted by applying information-theoretical methods.

The quantity which denotes the relationship between two pieces of information is termed the *transinformation* or the mutual information. The distribution $P(x,y)$ enters into this quantity; it gives the strength of the statistical correlation between two data sets:

$$I(X, Y) = \sum_{x,y} P(x, y) \mathrm{Log}\left[\frac{P(x, y)}{P(x)P(y)}\right].$$

The mutual information is found by summation over the random variables x and y on the right-hand side, and it characterizes the datasets X and Y. If the transinformation is zero, then the two datasets are completely independent of each other. When the transinformation is large, the two datasets are strongly correlated with each other. Making use of the transinformation, relationships between datasets can be established, such as for example languages or gene sequences. The origin of a text can be determined in this way. Hierarchies of relationships can be set up by defining a distance measure. In biology, one speaks of *phylogenetic trees*, in which various sequences of genes can be displayed.

[4]MacKay [2].

Joachim Hagenauer has reported[5] that his research group investigated the authorship of 85 articles which appeared in the New York press two hundred years ago, and had been ascribed to three different authors. Hagenauer's group succeeded in attributing twelve articles whose author was previously unknown to one particular author, namely James Madison.

Optimists in the data industry extrapolate such success stories into the future and claim that when the analysis of extremely large datasets containing petabytes (10^{15} Bytes) or more becomes possible, then the end of all theoretical analysis will be at hand.[6] They endorse the hypothesis that every theory is simply a model with a limited region of applicability. It would then be preferable to analyze large amounts of data systematically instead of constructing models which contain subjective prejudices. According to these "machine-thinkers", semantic analysis will also no longer be necessary, i.e. content will become unimportant. The large amount of data will make it possible by itself to discover a manifold of correlations and thus to find patterns which organize the data. As an example, they cite the genetic analysis carried out by John Craig Venter; using computer algorithms, he combined snippets of genes into a uniquely-determined gene sequence. In this method, samples of representatives of the same gene are cut up in different ways. On recombining, the resulting gene sequences are required to agree with each other. In medicine, one may face the problem that a new image must be projected onto the same coordinate system as a reference image already at hand. The new image is deformed, shifted and rotated until the mutual information between the reference image and the new image is maximized. We can expect that with increasing computational power, the potential for purely machine-controlled analyses will grow. Programs will have to be developed for this. In my opinion, however, it is equally important to develop abstracting models which can lead to generalized theoretical insights that go beyond the particular dataset underlying the specific model.

The method of testing hypotheses by means of statistics is well established. In order to test a coin to see whether it is fair, i.e. whether heads and tails occur with the same frequency, one measures the occurrence of heads and tails. They should occur with the same frequency, $h = ½$.

For a mathematical investigation of this question,[7] the *normal distribution* (Fig. 2.2) is utilized, as we recall from Mediocristan, the land of normality (cf. Sect. 2.2). The relevant bell curve has a maximum at ½ and a width which depends on the number n of tosses of the coin as $1/\sqrt{n}$. For example, for $n = 10{,}000$ tosses, the width corresponds to 0.01. Mathematically, 95 % of the area under the bell curve lies within the interval [0.49; 0.51] around the expected value of maximal frequency of occurrence (the maximum of the bell curve, here 0.50), and 5 % lies outside this interval. If after 10,000 tosses, heads has occurred 5200 times, then the associated frequency of occurrence, $h = 0.52$, lies outside the interval, and one can say with an error margin of 5 % that the coin is not fair. In contrast, if

[5]Hagenauer [3].

[6]Andersen [4].

[7]Cf. Bronstein et al. [5].

heads occurs 5070 times, i.e. with an occurrence frequency of 0.507, then one cannot refute the hypothesis that the coin is fair. The margin of error is again 5 %.

Such tests of hypotheses must be distinguished from *inference*, the method of consistent conclusion starting from an uncertain initial situation. This latter method is seldom or never taught in the education of a physicist or a physician. The Bayesian statistical theory on which it is based is controversial, since it includes prejudices explicitly in the probability calculus. In high-energy physics, the method is seldom used. However, in the analysis of the cosmic microwave background radiation and the associated cosmological models, Bayesian statistics is often employed. David John Cameron MacKay describes an introductory example which characterizes the problem well: "Let 10 distances $\{x_1, x_2, x_3, \dots x_{10}\}$ be given at which the decay of a radioactive source is to be measured. Because of the limited acceptance of the detectors, only values measured between 1 and 20 cm can be registered. How large is the decay length λ of the source?"[8]

The well-known exponential decay law determines the probability with which a data point will be measured at x with a given decay length λ to be

$$P(x|\lambda) \propto e^{-x/\lambda}.$$

This conditional probability must be normalized by the acceptance range of the detectors, which allows values only between 1 and 20 cm. From the Bayesian theorem, the conditional probability $P(x|\lambda)$ can be converted into the probability $P(\lambda|x_1, x_2, x_3, \dots x_{10})$ of measuring the parameter λ at given points x_1, x_2, x_3,... x_{10}. For ten measurement points which are sufficiently far apart, a curve is obtained that exhibits a clear-cut maximum at a particular value of λ. One obtains this result multiplied by the a priori probability distribution $P(\lambda)$, which specifies what we had assumed before the experiment about the decay length. Usually, one chooses a flat distribution; but without this advance assumption (*prior*), the Bayesian theorem cannot be applied.

Inference describes a method of going from a natural phenomenon to its theoretical description with parameters. The English scholastic William of Ockham (around 1300 A.D.) introduced the concept of ontological simplicity or parsimony in order to distinguish theories which are reduced to their essentials (*Ockham's razor*). The number of entities, according to Ockham, should not be increased beyond what is necessary. Therefore, physicists prefer a theory with fewer parameters over a more extended theory, even when the latter can make predictions over a wider range of data; the associated normalized probability will have a smaller value than that of the simpler theory (with fewer parameters). If the simpler theory in addition agrees well with the major part of the data, it is always to be preferred.

---

[8]David John Cameron MacKay: *Information Theory, Inference and Learning Algorithms*, *ibid.*, p. 48.

One has to differentiate between *inference* and *decision*. Inference is useful also in the humanities. Decision theory, in contrast, is mainly applicable to practical problems in the engineering sciences and economics, where one is trying to maximize profits or minimize losses, i.e. to optimize the results of a decision. We will treat decision theory in more detail in Chap. 4.

In order to arrive at the correct description of a random phenomenon, one has to take the *symmetries* of the problem into account. I wish to give an example of this: In an arbitrary textbook of physics,[9] one finds a page with the usual physical constants such as the velocity of light, the gravitational constant and Planck's quantum of action. What is the probability that the first digit of these constants is in each case a *one*? In principle, these numbers are completely random.

I find for example the velocity of light c = 299,792 km/s, the gravitational constant $G = 6.6 \times 10^{-11}\ \mathrm{m}^3\ \mathrm{kg}^{-1}\ \mathrm{s}^{-2}$, Planck's constant and many others. The numerical values of the constants are determined by the choice of units; one could have just as well expressed the velocity of light in inches/hour. What distinguishes this random distribution from the distribution that would be obtained if I chose points along a line which is divided into ten equal intervals? Is the above problem characterized by a special symmetry? If there is a law, then I am not allowed to change the order of the first digits by making an arbitrary choice of unit systems. If the physical constants are multiplied by factors, so that the first digit becomes one, the relative frequency of occurrence must remain the same. This means that the random distribution should be plotted on a logarithmic scale. The interval between the numbers one and two then has the following relative frequency of occurrence in comparison to the overall interval:

$$\frac{\mathrm{Log}(2) - \mathrm{Log}(1)}{\mathrm{Log}(10) - \mathrm{Log}(1)} = 0.3.$$

One can readily convince oneself that the above expression remains the same when all the arguments are multiplied by an arbitrary number that would result from a different choice of units. The table in the textbook cited shows a frequency of occurrence for the first digit *one* which lies somewhat higher. Indeed, the number *one* is found in 8 of 17 cases as the first digit.

As we have already seen in the discussion of the role of chance in theory (Sect. 2.1), symmetries limit random distributions. In spite of complete indefiniteness, symmetry considerations—in the case considered above the scaling invariance—allow a prediction which at first seems impossible. The concept of information introduced by Claude Elwood Shannon has far-reaching theoretical consequences, which I have attempted to sketch in this brief section. Together with the technological possibilities of high-speed computing, important resources for data analysis have been achieved. One of the most exciting developments in recent years is the application of information theory to quantum mechanics.

[9]Grupen [6].

- Indefiniteness of the first type is based on lack of knowledge;
- It can be reduced by acquiring more information;
- The potential information content is greater, the more improbable the result;
- Transinformation, inference and symmetries reduce indefiniteness.

3.2 Quantum Information

The fundamental choice between "yes" and "no" is the basis of classical digital information theory. All the logical operations and thus all computations can be carried out using the binary unit, the *bit*. In *quantum computing*, there is a different fundamental unit of information, namely the *quantum bit* or Qbit. The Qbit, in contrast to the bit, has infinitely many possible settings. This opens up many new applications and algorithms which cannot be implemented in classical digital information theory. The Qbit is composed of a superposition of two different quantum states; this superposition forms a new structure. We denote the two basis states by

$$|0\rangle = \begin{pmatrix} 1 \\ 0 \end{pmatrix} \quad \text{and} \quad |1\rangle = \begin{pmatrix} 0 \\ 1 \end{pmatrix}$$

In the double-slit experiment, this basis consists of the two states in which the electron passes either through the upper or through the lower slit. A physically somewhat simpler realization of a Qbit is obtained by superposing the angular-momentum states of an atom whose spin can be oriented upwards, $|0\rangle$, or downwards, $|1\rangle$.

The generalized Qbit state is a *superposition* of the two basis states $|0\rangle$ and $|1\rangle$. These states may be *entangled*; this expresses the fact that they can form a new unit. Mathematically, this entanglement is obtained by adding the basis states with complex coefficients:

$$c_0|0\rangle + c_1|0\rangle$$

In the quantum regime, the spin can be simultaneously directed upwards and downwards; only after a measurement is this quantum-mechanical indeterminacy removed, and a sharply-defined spin becomes associated with the system.

The sum of the probabilities for finding the 'spin up' and the 'spin down' states must be equal to one; therefore, the complex coefficients c_0 and c_1 are limited in range. Superpositions represent the same physical states when they differ only through multiplication by a complex number. The space of all possible states is thus a three-dimensional sphere with the state $|0\rangle$ pointing towards its north pole and the state $|1\rangle$ pointing towards its south pole,

$$\cos(\theta/2)|0\rangle + \sin(\theta/2)e^{i\varphi}|0\rangle$$

The angle θ denotes the angle between the spin direction and the axis through the north and south poles (usually termed the z axis); the angle φ denotes rotation along the equator (in the x-y plane). The 'yes' or 'no' orientations corresponding to a classical bit are given by the north pole and the south pole, respectively (see Fig. 3.2). The Qbit, however, can also take on all the possible orientations between the poles, i.e. in principle infinitely many possible orientations on the sphere. A measurement of the state can be carried out using an external magnetic field, which projects the spin onto one of the two basis states. The state $|0\rangle$ with 'spin up' will then be observed with the probability $\cos^2(\theta/2)$, and the state $|1\rangle$ with 'spin down' with the probability $\sin^2(\theta/2)$. The sum of these probabilities adds up to one, as required.

Combinations of Qbits correspond to more complicated states, e.g. the four states $|00\rangle$, $|01\rangle$, $|11\rangle$, $|10\rangle$ in the case of two Qbits. The generalized state with n Qbits is characterized by a sum of 2^n complex numbers, c_1, c_2, c_3, ... c_{2n} multiplied by basis states of the type $|000001010\rangle$. A quantum algorithm accepts an input of n Qbits and produces an output of n Qbits. This corresponds mathematically to a multiplication of the elementary basis vectors by unitary 2×2 matrices. Unitary matrices conserve probability; the individual spins are found on measurement to be directed either 'up' or 'down'.

An important application of quantum information is *encryption*: How can one invent a code that is very hard to break? Classical methods include transposition, in which the order of the characters is changed in the message to be sent, or

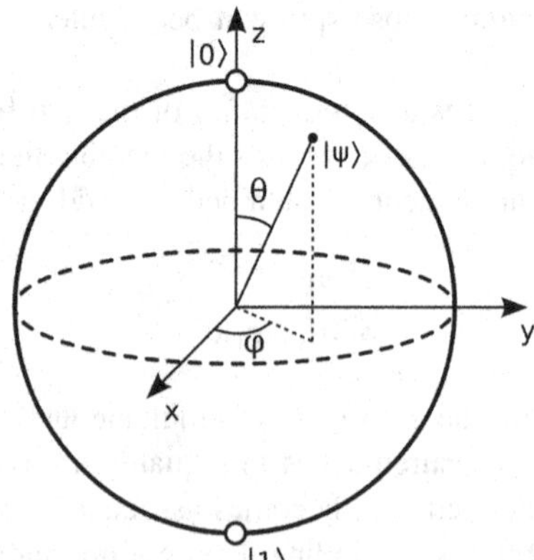

Fig. 3.2 Representation of a Qbit, the basic unit of quantum information and quantum computing. Unlike just the two classical states $|0\rangle$ (north pole) and $|1\rangle$ (south pole), the quantum state can take on all orientations on the so-called Bloch sphere. (Image licensed under: Attribution-ShareAlike 3.0 Unported (CC BY-SA 3.0).)

substitution, whereby every letter in the alphabet is replaced by a different letter. Both methods however can be decoded by analyzing the frequency of occurrence of the characters. Gilbert Vernam discovered an encryption method which is secure.

In his method, the characters in the original message are translated into a binary alphabet: the letter a becomes 0, b becomes 1, c becomes 10, etc. Then a random key is generated which consists of just as many zeroes and ones as the translated message. The message's sender adds this random key to the binary original text, carrying out the addition modulo 2 ($0 + 0 = 0, 0 + 1 = 1, 1 + 1 = 0, \ldots$). The coded message is then transmitted to the recipient, who also has possession of the random key. The recipient decodes the message by again adding the key, since a double addition restores the original binary message. For example, a 0 in the message becomes again 0, since $0 + 0 + 0 = 0$ and $0 + 1 + 1 = 0$. The random key can be re-used many times, because the sum of two coded messages is equal to the sum of the plain texts. The chief difficulty of this method is the secure transmission of the key from the sender to the receiver. Here, quantum physics can be of help.

What is new in quantum cryptography? Its task is to send the key securely from the sender to the receiver. To this end, the sender makes use of particles with basis states $|0\rangle$ and $|1\rangle$. Sender and receiver make it openly public that the spins of the particles will be polarized and analyzed either along the z axis or along the x axis. The corresponding operators can be denoted as O_z and O_x. It is essential for cryptography that these operators do not commute:

$$O_x O_z - O_z O_x \neq 0.$$

The sender prepares a random distribution of particles polarized in the z- or the x-direction. The receiver then analyzes the stream of particles with his measurement apparatus either along the z- or the x-direction without knowledge of the prepared distribution. On average, he will thus correctly identify the states of only half the particles. In the third part of the procedure, sender and receiver make their preparation and analysis of the particles public. They ascertain for which particles their results for the states $|0\rangle$ and $|1\rangle$ are in agreement. These values of 0 and 1 are kept as their code key.

The security of this key is based on the "no-cloning theorem"[10] of quantum computing, which states that an unknown quantum state cannot be copied. Quantum cryptography has been tested in practice: In October, 2007, the method was successfully used during the Swiss parliamentary elections. The Canton of Geneva sent the election results securely from the counting location to the central election headquarters. The method makes use of two channels: the classical radio channel over which the information is exchanged, and the quantum channel using light (i.e. light particles or photons) over which the quantum key is sent.

The inverse problem in quantum theory is "decoding". How can one decrypt the indefinite original text when only the translated version is known, and the key is not? Physicists are often confronted with similar problems. From experiments, for example, scattering data are obtained which result from collisions of two reaction partners a and b ('projectile' and 'target') over a range of energies. The physicist is

[10]Wootters and Zurek [7].

seeking the interaction between a and b, which must be known in order to predict other data. The strong nucleon-nucleon interaction was investigated in great detail by such scattering experiments. It depends upon the spins and angular momenta of the collision partners and is different for charged and neutral nucleons, i.e. for proton-proton and proton-neutron scattering. Once the interaction has been determined from the scattering experiments, this same interaction can be used to calculate systematically the structure of atomic nuclei, ranging from the very smallest nuclei such as alpha particles, up to large nuclei like that of uranium.

In the complicated case of nucleon-nucleon scattering, the scattering potential cannot be uniquely determined. However, in the simpler case of electron-nuclear scattering, it is possible to reconstruct the electromagnetic interaction potential from the scattering data, and thus determine the shape of the atomic target nucleus.

Since the quantum-mechanical Schrödinger equation is a differential equation similar to the eigenvalue equation for the musical notes produced by a drum, the question posed by Mark Kac in 1966, "Can one hear the shape of a drum?",[11] is relevant in this connection. The frequency spectrum of the drum depends upon its shape. Does the spectrum contain sufficient information to reconstruct the unknown shape? This problem was finally answered in the negative in 1992 by Carolyn Gordon, David Webb and Scott Wolpert, who ascertained that in two dimensions, there are different surfaces which can produce the same acoustic spectra.[12]

Is there hope that the development of quantum information and quantum computing will also deepen and elucidate our understanding of the interpretation of quantum mechanics? For some time, the school of Carl Friedrich von Weizsäcker has emphasized the importance of the concept of information for quantum mechanics. Holger Lyre deliberates in his book "*Quantentheorie der Information*" on the connection between information theory and quantum theory. Matter and energy are interpreted as manifestations of quantum information. Does the similarity of the Bloch sphere to three dimensional space indicate that space has to be three dimensional? Information is considered to be the primordial substance of the universe.[13] Elementary chance alternatives—"does the detector 'click' or not?"—are the elements upon which our knowledge is based. Are they also the elementary components of the universe? The concept of vagueness, which originated in semantics, could then be expanded and extended to objects. This idea has also already been discussed by the philosophers.[14] Are clouds vague objects? Which water droplets belong to a particular cloud? All clouds seem to have similar fractal structures, which means that their shapes repeat over and over on different length scales. Probably it must be decided from case to case which laws apply to such "vague" objects before one can make any general statements about them.

[11]Kac [8].

[12]Gordon et al. [9].

[13]Cf. Wheeler [10].

[14]Cf. Unger [11].

Are there unsharp, fuzzy objects, or is the world completely described in terms of facts? Ludwig Wittgenstein asserts that "The world is everything that is the case".[15] Vague statements are in this view meaningless, since they are located somewhere between these facts and cannot be specified to be true or false.

Anton Zeilinger, in contrast, argues from the viewpoint of quantum information: "In quantum mechanics, we cannot make statements about what is the case, but only statements about what could be the case".[16] He means that the quantum-mechanical state describes the system with all possible consequences, of which some are realized in practice, but only after measurement. The quantum regime is thus a world of possibilities.

Quantum information theory can open this world to us, for example through the construction of a quantum computer, which would make use of new computational methods that could solve some complicated problems rapidly. The idea of the quantum computer is based on the physical implementation of entangled states (see the beginning of Sect. 3.2), with which computational operations can be carried out. These states of quantum objects are however extremely sensitive to disturbances from their surroundings, which in general destroy the phase relations between the basis states $|0\rangle$ and $|1\rangle$ (*decoherence*). Therefore, it is difficult to build such a computer. In biological systems, decoherence is on the other hand often useful. The noise from the surroundings of the photosystem makes energy transport[17] in photosynthesis more efficient. In general, thermal fluctuations are more important than quantum fluctuations in biology, owing to the high temperatures (300 K).

The possible states which are coded in the quantum-theoretical wavefunction should be clearly distinguished from other "possibilities" or other "possible worlds", of which I will mention just three here. The *many-worlds interpretation*[18] of quantum mechanics asserts that after each measurement, different, parallel worlds exist side by side. Even if the state $|0\rangle$ is measured in our world, there is another world in which the state $|1\rangle$ continues to exist, but we cannot communicate with it. In the *multiverse theory*,[19] there are many universes, among which our universe is just one of those possible. It could be—but need not be—that in the multiverse, the many-worlds interpretation is indeed realized, since every quantum-mechanical state evolves in any case into many worlds.[20] A multiplicity of *possible worlds*[21] has also been discussed by philosophers. The philosophical considerations have developed concepts of modal logic, which deal with questions of the possible and the necessary.

[15]Wittgenstein [12].

[16]Zeilinger [13].

[17]Chin et al. [14].

[18]Everett [15].

[19]Susskind [16].

[20]Aguirre et al. [17].

[21]Lewis [18].

The wavefunction remains an important information carrier in quantum mechanics, and it contains all the possible results of later measurements. In comparison to the other theories mentioned above, it appears quite realistic.

- Qbits are *superpositions* of a 'yes' and a 'no' decision;
- Quantum coding makes secure encryption possible;
- Quantum experiments: Can we hear the shape of a drum?
- The quantum regime is a world of possibilities.

3.3 Complexity

Descartes has given us precise instructions for a universal mathematics: "Method consists entirely in the order and disposition of the objects towards which our mental vision must be directed if we would find out any truth. We shall comply with it exactly if we reduce involved and obscure propositions step by step to those that are simpler, and then starting with the intuitive apprehension of all those that are absolutely simple, attempt to ascend to the knowledge of all others by precisely similar steps".[22] We started with the simplest concepts of information and we now wish to discuss the algorithmic concept of *complexity*, which is closely related to Shannon's definition of information. For the information dynamics that we wish to derive, complexity and indefiniteness are of central importance. If we acquire a quantitative understanding of these two concepts, we can calculate the gain in information. The goal of this section is to delve into the general concept of complexity.

What does complexity mean? "Neither the philosophical discussion nor the systems-analytical treatment has up to now yielded a satisfactory answer to this question. Older conceptual descriptions, which explain 'complex' as a composite, are not sufficient".[23] In a purely algorithmic sense, Andrei Nikolajewitsch Kolmogorow and Gregory John Chaitin have defined the complexity of a character string in terms of the shortest program which can generate that string.[24] The character string "01 01 01 01 01" is simpler than the string "01 00 10 11 01". For the first sequence of numbers, there is a short program which consists of only two lines of code; the first line contains the block 01 and the second line the instruction to repeat this block five times. Thus the first character string is not very complex. A program for the second string would have to list the individual zeroes and ones in sequence. This program is therefore just as long as the string itself,

[22]Descartes [19].

[23]Luhmann [20].

[24]Li and Vitányi [21].

i.e. the complexity of this string is maximal. This concept of complexity is more a measure of the intrinsic randomness of the string than of its degree of organization.

Charles H. Bennett attempted to correct this defect by defining the "logical depth" of an object as a measure of its self-organization.[25] The logical depth denotes how long a program would have to be which could logically reconstruct the process of emergence of a system. I would rather stay closer to the idea which forms the basis of many definitions and which, simply put, states that: The more complex something is, the more that can be said about it. With this I mean not the detailed description of the character string itself, which can be rather lengthy even for simple random distributions, but rather the description of its overall properties. Here are a few examples:

A sequence of randomly-distributed ones and zeroes can be completely described by the statement that the probability for the occurrence of each 'one' is 50 %. Making use of the laws of probability, one need not take up the special structure of a particular string of ones and zeroes; that would indeed require a program just as long as the string itself. Instead, one can classify its complexity through the complexity of the model. Let us compare a random sequence with the string of characters that occurs in a text; this allows us to say much more about the text. One can discuss the choice of words, the style and the story line. Here, different levels of description are possible: a simple analysis of the frequency of occurrence; an ethnological-linguistic analysis as would be carried out by someone visiting a foreign culture; a classification of the story and the characters in the text within a narrative consisting of a temporal ordering of individual events, etc.

A *statistical complexity* has been defined which consists of the product of the entropy times the system's disequilibrium, i.e. its deviation from equilibrium.[26] The fully random and the fully ordered systems have zero complexity. This is a meaningful definition for finite systems. In order to penetrate further into the concept of complexity, we require a definition of a "complex" system. A *system* refers in general to an organized collection of elements into something larger, which thereby acquires new properties. The associated systems theory has evolved since the 1940s from a mixture of biology, mathematics and computer science. Its holistic approach distinguishes this theory from the traditionally reductionistic approaches to understanding Nature. While the Cartesian-reductionistic method has been historically successful in physical research, the importance of collective processes has grown with the development of new materials and the investigation of biological systems. As 'system', usually an open system is considered, which is embedded in an environment that interacts with it. A typical example is the ecosystem of a pond.[27,28]

---

[25]Bennett [22].

[26]Lopez-Ruiz et al. [23].

[27]Odum [24].

[28]Cf. Mainzer [25].

Modern considerations and descriptions of complexity often contain only a listing of the parts of this rapidly-developing field: The pond indeed has a clearly-defined geometrical boundary with its environment, but it is in a process of continuous exchange with the surroundings. It is complex because it hosts a great variety of lifeforms whose lives are also influenced by the environment, i.e. the insolation, oxygen exchange and possibly inflows into the pond. Formally, one could say that the "definite" system derives "information" from its "indefinite" environment. The complexity of the system determines the state of the system. Matter, life, the unity of brain and body, computer algorithms, simulations, the economy, and finally culture and society are all subsumed under the aspect "complexity". This complexity is often seen as an evolving complexity. The emergence of structure in each of the subfields listed above can be better understood through studies of their complex and non-linearly interacting parts. The lexical representation characterizes the current level of research, which is evolving rapidly and with a diversity that is difficult to penetrate.

Here, I would plead for a moderate and simplified approach. Let us initially leave aside the temporal evolution of the complex system and look at the system itself. Are there ways to describe its properties and its structure non-specifically? Are there generic properties which could be applied to every system in the above list?

A complex system consists of active elements ("agents") and interconnections ("relations"). The organization of the interconnections is woven together in a complicated manner, intertwined and entangled. Even if there are no nonlinearities on a microscopic time scale, the system can react upon itself through loops and thereby produce nonlinearities over a longer time scale. Since for biological systems, we cannot regress to an arbitrarily short time scale, we have to allow for such effects from the beginning. The strength of the interconnections varies, and effects can proceed in both directions along them. The separation of the agents from their relations is not generally apparent; it demands clarification.

For the elementary particles in microscopic physics, for example, we distinguish between *matter particles*, i.e. quarks and leptons, and the *force-carrying particles*, i.e. gluons, photons and the gauge bosons which mediate the strong, the electromagnetic and the weak forces between the matter particles. The net number of matter particles, that is the number of particles minus the number of antiparticles, is limited by conservation laws for the total electric charge and the baryon number (one-third of the number of quarks). Bosonic force carriers have no such limits. In the Standard Model of elementary particles, such a separation between the agents (matter particles) and the interconnections (forces) is therefore feasible. "There is no matter"[29] is a misleading sentence. It is true when 'matter' is meant in the old sense of "immutability"; for there is no immutability in quantum mechanics. The sentence is false since the net number of particles minus the number of antiparticles remains constant. Photons can, for example, create only pairs of particles consisting of an

[29]Cf. Dürr and Zimmer [26].

electron and a positron, its antiparticle, simultaneously. The question arises as to whether interconnections are in principle simple or whether they interact among themselves. The gluons, for example, which hold the quarks together in the nucleons, do interact with each other. Do interconnections act in spacetime, or in a functional space of their own? Physicists prefer the former model, biologists show a tendency to prefer the latter. In neuro-physiological models, the neurons are the agents and the synaptic couplings are the relations (interconnections). The neuronal network abstracts from the spatial arrangement and models the essential connections.

In general, a complex system can be divided up into various hierarchies, e.g. organic matter and inorganic matter. These hierarchies form the horizontal structures of the system and themselves are again subdivided into configurations. The inorganic matter can occur e.g. in a liquid, solid or gaseous phase. It is important to consider these horizontal levels; in biology for example the horizontal gene exchange.[30] In the area of microbial biology, the horizontal exchange is responsible for the development of resistance to antibiotics. In the evolution of species, a gradual improvement through horizontal exchange of information attained a high level of complexity before a transition to a new hierarchy became necessary. Nigel Goldenfeld and Carl Woese argue that the old classification of single-celled organisms into prokaryotes (cells without nuclei) and eukaryotes (cells with nuclei) should be replaced by a classification into bacteria, archaea and eukaryotes. This should be based on genetic similarities of their RNA. The exchange of genes between these organisms played an important role in the early evolution of life, according to Carl Woese.

Finite temperatures allow a system to fluctuate between different configurations, which are accompanied by quite diverse interactions. For a protein, for example, there are many different possible foldings, all of which have energies in the neighborhood of the overall energy minimum. A minor change in the chemical structure of the protein or in the solvent can totally change the protein's folding.[31] It is therefore not possible to predict the form of larger proteins from calculations. The goal of calculations must thus be modified; it is more reasonable to compute probabilities for the various configurations. The complexity is then given by the distribution of the interactions which are produced by those various configurations.

Sandra Mitchell, under the title "Complexities. Why we are just beginning to understand the world",[32] listed the following characteristics of the new paradigm for acquiring knowledge: There is a plurality of explanations on many planes of explication. This is complemented by the pragmatism of finding the best possible answer to a particular question. Instead of Descartes' universal mathematics, she prefers a dynamics of knowledge, which encompasses theoretical research, experimental investigations, and practical activities combined into a new pattern.

[30]Cf. Goldenfeld and Woese [27].

[31]Cf. Parisi [28].

[32]Mitchell [29]. See also Mitchell [30].

An isolated system has no coupling to the outside world. Let us assume that we know all of those parameters which characterize the interconnections among themselves. Then the system could be readily evolved in time. Its internal dynamics would have to be completely describable in terms of the inner forces of the system. The time scales of the internal system would be known. Chaotic behavior of the internal dynamics could occur, and the associated exponents which characterize the divergence of the trajectories could be determined. Their complex behavior is determined generically. Even in a limited study of complexity on a certain level, the interactions of the system with its environment must still be taken into account. This is fundamental for open systems which are not in equilibrium. How is it possible to separate a system from its environment? All the undetermined external influences on the system can be attributed to the environment. The boundary between system and environment is therefore variable. Those parts of the environment which are well understood can be included within the system. The environment changes the system, since the latter responds to external influences, which makes it seem as though internal interaction parameters have acquired a stochastic, i.e. random, character. Nevertheless, it is structurally important to separate the system from its environment. Distinguishing the determinacy of a system from the indeterminacy of the surrounding world is in my opinion an important feature of Luhmann's systems theory, which should be included as a rule for the modeling of complex systems. "The theory begins with a difference, with the difference between system and environment, insofar as it claims to be a systems theory; if it is another theory, then it must be based on another difference. It begins not with a unity, with a cosmology, with a concept of the world, or with a concept of being or something similar, but rather it begins with the difference".[33]

What happens when we couple the system to its environment? The system will be driven to follow the external influences; depending on its internal dynamics it will enter into an interrelation with the external forces. Resonances may amplify the externally driven motions. We are familiar with this behavior from machines which are controlled externally. The indefiniteness enters as a shaping element through forces from outside the system. Is the temporal evolution of the complex system defined by infinitesimal steps in the direction towards an increase in the complexity of the system itself and a reduction of its indefiniteness? How does one quantify the coupling of the system to its environment?

Indefiniteness represents an important quantity for the calculation of the gain of the system through its acquisition of information. The actual information which is added to the system increases the system's complexity and reduces the indefiniteness of its environment. The specific change of the complexity relative to the change in the indefiniteness can be a measure of the gain in information, as I shall show in the following section. In Sect. 2.4, on the semantic aspects of indefiniteness, the indefiniteness was attributed to each of the experts. It is defined by their opinion functions, and is thus not objective. Similarly, the degree of organization,

[33]Luhmann [31].

i.e. the complexity of each information participant, varies individually. Indeed, the gain in additional information will be different for the different information participants, depending on their intrinsic complexities. By averaging over larger entities, such differences can be eliminated, but it seems to me to be impossible to remove them entirely from the calculation.

Luhmann assesses the complex world skeptically: "By system differentiation, science thus anchors its function of acquiring new knowledge in the world as it is seen. But that also means that the distance between knowledge and what is worth knowing continually widens, and science can no longer discern a goal of knowledge which, if it would be attained, would bring it to repose. The system can no longer comprehend itself in a teleological manner, but only autopoietically: as an activity which continually feeds upon itself. Science thus becomes the means through which society renders the world uncontrollable".[34] This sounds paradoxical: Normally we would assume that science makes the world more understandable and predictable. But the above quote emphasizes the gap which exists between scientific goals and acquired knowledge. The established methods, the data obtained, and their understanding multiply the hypothetically possible new results, thereby never coming to a standstill. This creation of complexity seems limited only by economic factors and human potential.

- Complex systems show a high degree of organization;
- Hierarchical planes and horizontal structures form the fundamental units;
- Differentiating between the system and its environment is decisive;
- The system is definite, its environment is indefinite.

3.4 Obtaining Meaning from the Information Cloud

Information technology has had a decisive influence on the second half of the 20th century. The number and computing power of computers has increased explosively; at the same time, their processors have become smaller and memory capacity larger. Moore's Law predicts that the number of switching elements per unit area will double every 18–24 months, so that computing power grows exponentially.[35] This has indeed occurred. The programming and logistics of computational operations has likewise made impressive progress. What does this rapid development of information technology mean for knowledge? Has a new path towards determining the essentials of information transfer been opened? This section contains my central

[34]Luhmann [32].

[35]Cf. Moore [33].

thesis, which maintains that the reduction of indefiniteness in comparison to the increase of complexity specifies the *value* of information.

The pioneering achievements of the past decades undoubtedly include the Internet and the World Wide Web. In 1989 at CERN,[36] Tim Berners-Lee laid the cornerstone for the World Wide Web in his project proposal for "Information Management". In the introduction to his proposal, he wrote: "This proposal concerns the management of general information about accelerators and experiments at CERN. It discusses the problems of loss of information about complex evolving systems and derives a solution based on a distributed hypertext system".[37] Many staff members of the European accelerator laboratory CERN change their place of residence every 2 years, that is they are sent to CERN for a certain time and then return to their home laboratories all over the world. These staff members have to quickly familiarize themselves with the new situation, in order to optimally continue the work carried out by their predecessors. If CERN were a static organization, one could write down all the necessary information in a large book. But since the experiments at CERN evolve dynamically, and often a whole series of details of the detectors and the accelerators change from time to time, they cannot be effectively documented in a thick book and be kept up to date. Even if such a logbook were kept for all the participants in a particular experiment (and they can number in the hundreds), one would never find the relevant place in the book where a particular detail was described. For that reason, Berners-Lee developed the Hypertext Markup Language (HTML) to structure the manifold content that can consist of texts, images and links to other HTML documents. This opens the possibility of cross-linking into a vast network. This project marked the beginning of one of the most venturesome chapters in the new information era. One could say that the World Wide Web combines and assembles collective knowledge, contained in a mega-memory and corresponding to an immense library. Embedded in this network are encyclopedias such as Wikipedia which provide rapid look-up of keywords. Its inventory of information content is continually growing.

Aside from the scientific utilization of the Internet, there are a number of commercial enterprises which make a profit from the distribution, channeling and provision of information. I refer to Internet search engines such as Google, Bing, Yahoo, iTunes and others. It is not astounding that many people search for information in order to find their way through the consumer jungle. Google of course serves not only this commercial aspect; with Google Scholar and the attempt to establish a world library in digital form, their own vision of contributing to the well-being of humanity through the spreading of information becomes apparent. One of the founders of Google, Sergey Brin, has said: "Knowledge is always a good thing,

[36]CERN is the abbreviation for "*Conseil Européen pour la Recherche Nucléaire*" and is the name of the European accelerator center in Geneva.

[37]Cf. Berners-Lee [34].

more of it should be shared." His colleague Larry Page on the Google board of directors states that "Solving big problems is easier than solving little problems".[38]

The Google developers began their project with this optimism, and it seems to me that they can be seen within the tradition of the authors who contributed to the "*Encyclopédie ou Dictionnaire raisonné des Sciences, des Arts et des Métiers*" in 18th-century France. Voluntarily or involuntarily, the google system fulfils the same mission: "In articles in which the watchdogs might suspect potentially objectionable ideas, such as 'The Soul', 'Free will', 'Immortality', or 'Christianity', the encylopedists repeated the orthodox teachings, while in quite different articles, where no one would suspect such controversial topics, diametrically opposite principles were divulged with a plethora of arguments and at the same time hidden clues which the insiders among the readers rapidly learned to interpret and to interconnect".[39] Sergey Brin and his colleague Larry Page want to open up the specialized scientific knowledge in medicine, biology and other natural sciences, which has become practically inaccessible to the general public, to a wide spectrum of users. In this way, they will set a new realignment of thinking in motion, whether or not that was their original intention. The Google search engine provides the desired information, together with advertising which aims at selling something. Other developments such as Google Street View and the storage of users' IP addresses[40] shed a less favorable light on Google's methods, since they show an increasing lack of respect for the privacy of individual users.

Applications such as commonly-used editing programs should be stored in a data cloud which is available to everyone, everywhere. Meaningful analyses with data comparisons require handling enormous amounts of data in the range of petabytes. The following list can give some idea of the orders of magnitude; it gives the amounts of data which can be stored on various media, increasing by factors of 1000: Megabytes can be stored on a diskette, gigabytes on a hard disk or flash memory, terabytes in high-capacity disk drives or flash memories, and petabytes are best stored in the data cloud, i.e. in decentralized storage servers which are accessible from everywhere via the Internet. The Web organizes information by comparing similarities in texts and indexes addresses as nodes of greater or lesser importance. But it does not analyze meaning. For that, a semantic Web is in planning which will attempt to analyze meanings using computers. Linking up information from different sources can lead to discoveries. Speech recognition and natural language translation are at the forefront of semantic web applications. In the library, when searching for a particular book, one often discovers by chance another important book. The extremely rapid online searches for literature and its evaluation render such lucky discoveries unlikely these days. However, the supporters of more and more internet linkage claim that even computers, "wandering around" rough the vast amounts of data in the net, will be able to sniff out important regularities

---

[38]*Enlightenment man*, in: The Economist, Technology Quarterly, Q4 2008.

[39]Cf. Friedell [35].

[40]http://www.ixquick.com/; *Ixquick* is a search engine which does not store users' IP addresses.

and new, previously unrecognized interrelations (*serendipity*). To be sure, I find it dubious whether this technology will pamper us with such happy coincidences via the back door.

In Sect. 3.1, we were interested in technical information transmission, with a sender as source, a transmission channel and a receiver; now we will turn to the analysis of the *content* and *meaning* of information. For the mathematical theory of Shannon's information, a source is the origin of randomly ordered characters which can be compared with other sources of likewise randomly ordered characters. This task can be carried out efficiently by computers and has attained a high level of perfection using modern machines. The data can be input in a straightforward manner and the information is transmitted in an optimized code with the highest rate of compression. Such mathematically treatable information is termed *syntactic information*. In contrast, when we speak of data and their meaning, we are referring to *semantic information*. At this point, we shall end our discussion of the purely technical aspects of information theory in order to learn more about this other kind of information.

What does "meaning" mean? In an article on "The Meaning of 'Meaning'", Hilary Putnam[41] distinguishes between the "extensional" and the "intensional" aspects of meaning. On the one hand, the word "cat" includes all cats in a well-defined set of cats with all the problematic cases which we discussed in referring to fuzzy sets. Here, we can think of the range ("*extension*") of the concept. On the other hand, the word "cat" has other meaningful content ("*intension*"). This content includes figurative meanings such as "scaredy-cat", someone who is fearful of everything. This second aspect depends on the person who is speaking; he is referring to something which goes beyond the set of cats. The first aspect is closely connected to the concept of truth; an object either belongs to a particular set, or it does not. The second aspect is part of the language usage of the person and of the segment of society which speaks that language. In the case of technical terms, that society can also be a group of experts who have ritualized and implemented a certain jargon. The non-expert has to trust that the usage of the concept is correct. The social context is thus very important in discussions of meaning. It provides a new frame of reference which is referred to by the concept. Fearful people run away from danger, like a cat that has been startled.

Frank J. Miller intended to provoke his readers when he published a commentary on information theory with the title "I = 0 (Information has no intrinsic meaning)".[42] He argues in that article that information is meaningless by its very nature. The same information can cause very different reactions in different recipients when their frames of reference are not the same. Depending on their interests, personal histories and emotional dispositions, various people interpret the same information in quite different ways. Successful communication is based upon knowing the recipient and on formulating the message in such a way that the

[41]Putnam [36].

[42]Miller [37].

information it carries will be understood correctly. This aspect of information was emphasized early on by Donald MacKay, who proposed summarizing the possible states of the recipient in a conditional probability matrix.[43] The transmitted meaning can then be defined as a special function of the message on the possible states in this matrix. The matrix characterizes the psycho-physical state of the recipient and can thus not readily be expressed in terms of numbers.

From the viewpoint of the receiver of information, it is very clear that he has indeed gained information if he now knows something which he previously did not know. i.e. when his state of knowledge has increased. I will attempt to analyze this process with the tools of systems analysis. The mathematical theory of information, as explained in Sect. 3.1, makes no statements about this process. The indefiniteness as defined by the receiver can best be used to characterize the process. This indefiniteness will be decreased when new information is correctly transmitted. In order for this information to become knowledge, it must be organized, processed and linked to other knowledge, i.e. the organization, the linking and the structuring of the knowledge by the receiver must also be embedded in the estimate of knowledge gain. The biblical parable of the word that can fall on stony ground with little earth, or on good ground,[44] describes this situation accurately. For someone who has received a humanistic education, it may be more difficult to deal with formulas and mathematical concepts. They are hard to integrate into the knowledge he already possesses, since there are few points of linkage. I assume that the system of available knowledge is defined by its complexity. It communicates with the indefinite environment in which it is embedded by means of information. Initially, the state of knowledge of the receiver has a certain, known complexity, while the still unprocessed elements of information which he proposes to acquire have a certain degree of indefiniteness.

When the receiver processes this information, the complexity of his knowledge base increases and the indefiniteness of the environment decreases. The information is integrated into and linked with the knowledge already present. Now, my hypothesis is that the relative ratio of the change in complexity to the change in indefiniteness provides a measure of the information gain; that is, when a small "portion" of information has a large effect, then that information has a high value or worth. It fits so to speak very well into the conceptual world of the receiver. It need not conform to his previous opinions, i.e. it need not add *directly* to his existing knowledge, but instead it increases the overall complexity of his knowledge by making new linkages possible. This *worth of information* can distinguish between valuable information and useless information, and it is thus different from Shannon's quantification of information, which is based on the surprise value of the information.

[43]Cf. MacKay [38].

[44]Parable in the New Testament, Mark 4, 1–8: https://www.biblegateway.com/passage/?search=Mark+4&version=KJV.

In Fig. 3.3a, I have attempted to illustrate this process in a purely schematic fashion. The circle represents the system of knowledge and contains elements of knowledge which are linked together. It is clearly delimited from its indefinite environment, which contains informational elements of various form and content. Figure 3.3b illustrates the action of acquisition of information, which introduces an indefinite element (triangle) into the system. Through this process, the original message (small triangle) gains in meaning (large triangle) by linking with the existing elements of knowledge in the system. The complexity of the system changes (increases). The relative change in the complexity of the system compared to the change in the indefiniteness of the environment determines the *Worth of INformation* (WIN). If the indefiniteness of the environment changes by only a little, while the complexity of the system increases strongly, then the information is very important for the development of the system and thus has a high value.

It must be verified from case to case just how great the specific change in the complexity is for a given change in the indefiniteness of the environment. The postulated law for the worth of information (WIN) has the following form:

$$\mathrm{WIN} = \Delta\mathrm{complexity}/\Delta\log[\mathrm{indefiniteness}].$$

The logarithm is a monotonic function which takes into account the fact that the disorder in the indefinite environment increases logarithmically, just as in

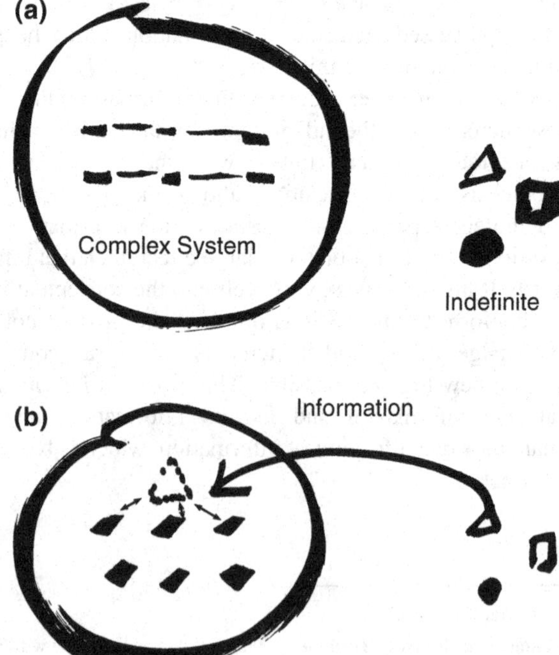

Fig. 3.3 **a** The complex knowledge system is separate from its environment containing indefinite elements. **b** Information transports a still-indefinite element from the environment into the knowledge system, which links the new element to the existing knowledge elements. This increases the complexity of the system. The indefiniteness of the environment decreases at the same time

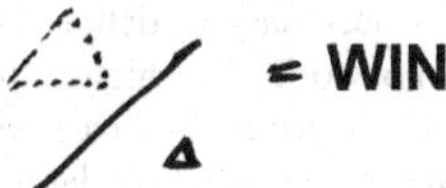

Fig. 3.4 The ratio of processed information (and resulting gain in complexity) to the change in the indefiniteness of the environment yields the Worth of INformation (WIN)

thermodynamics. In Fig. 3.4, the Worth of INformation (WIN) is illustrated graphically. If the information has a high value, the element which is introduced into the system is "enlarged" by linking with many existing components of the system.

The information cloud which is even now being constructed in the virtual space of the Web will doubtless increase the amount of useless information available. In the "Battle of the Clouds",[45] the services of the clouds are indeed free of charge, but useful information is delivered to us together with unimportant garbage. Making use of the WIN criterion, every user of the Internet could in principle establish a profile which filters only those elements out of the mass of available information that represent a real gain for him or her. It is annoying that the large service providers such as Google generate profiles of their users according to their own commercial criteria, while we ourselves have no access to the software which would allow us to evaluate and thus regulate the flood of information. According to Eli Pariser's study "The Filter Bubble",[46] Google has personalized Internet searches since 2009; that is, their algorithm suggests results which fit the Google profile of the user. In this manner, preconceived opinions are intensified instead of checking and structuring knowledge by the acquisition of new information. The information society desperately needs a method for evaluating the information which is provided to it.

As we saw in Sect. 3.1, Shannon's definition of information is closely related to the *entropy*. Thermodynamics makes use of the entropy in order to determine the quality of energy. The project of information dynamics is shaped along the same lines as thermodynamics, and it aims at determining the *quality of information*. We have today a situation which is analogous to that of thermodynamics in the 19th century. At that time, newly-constructed steam engines needed a criterion to establish how efficiently they used fuel to provide energy; Nicolas Léonard Sadi Carnot (1796–1832) derived such a criterion from his idealized theory of cyclic processes.

> Carnot prepared the way for the concept of entropy, which together with energy is a determining factor in our industrial economy. In simplified form, the entropy (S) of a quantity U of energy at the temperature T determines the quality of that energy, i.e. how much *free energy* ($F = U - TS$), which can perform useful work, is available within the energy U. The law of conservation of energy was not sufficient to permit further technical

[45]Cf. *Battle of the clouds*, in: *The Economist* October 17th (2009), p. 13.

[46]Cf. Pariser [39].

development of the engines; it was necessary to define the *free energy* as that part of the energy which can be converted into work. The higher the entropy S in a quantity U of energy, the less useful work can be performed by that energy. When cooling water from a reactor is fed into a river, the river then contains a large amount of energy; but it is no longer useful for performing work or powering sophisticated industrial installations.

When information is considered in relation to indefiniteness, the quantity of useful information (WIN) can be computed. This would represent a similar advance for information technology as did the concept of entropy for the energy economy.

Luhmann maintains that there is information only within the system. "It is thus pointless that masses of information are present in the environment and that their transfer into a system depends on the system itself; rather, the system reacts to its own states, to stimulation which it experiences itself, in order to convert them into information".[47] He distinguishes between the indefinite complexity and the definite complexity of the system.[48] This distinction is carried out differently in my approach, in which I identify the indefinite complexity with the indefinite environment. To be sure, the concept of definite complexity remains a challenge for me; its quantification is still not complete. In the Appendix, I show using the example of the logistic equation that a simple system can possess an intrinsic complexity. It is defined in that case by the characteristic constant a which determines whether the system will behave predictably in terms of periodically recurring events, or will enter the chaotic regime and thus become unpredictable. The indefinite environment of this system is simulated by a random noise term.

A theoretician would like to see a new experiment performed whenever it promises to deliver important information. But the experimenter will perhaps not agree with the opinion that the information thus obtained will generate more complexity. Sometimes it appears to be that way. But ideally, the experimenter hopes that the new results of his experiments will make everything simpler. To this we can reply: The system consists of data and theories about certain objects. New data yield new information which can, perhaps, simplify the theory. The overall complexity, however, will still increase, since this simpler theory can explain a greater range of phenomena. The complex system of knowledge must therefore be considered as a whole and not limited to just the theory.

In the quote from Heidegger's essay "The Principle of Reason" (see Sect. 2.6), the rationality of our thinking, which limits language to the role of carrier of information, is questioned: "Dare we, if it should be so, yield up this most memorable in favor of simply calculating thought and its enormous successes? Or is it not our responsibility to find paths on which our thinking can answer to the memorable, instead of sneaking past it, bewitched by calculating thought?".[49] It may seem paradoxical, but calculating thought can help us to separate out the

[47]Luhmann [40].

[48]Niklas Luhmann, *ibid.*, p. 301.

[49]Heidegger [41].

memorable and the non-trivial from the background noise of the increasingly loud and strident information orchestra.

- The WorldWide Web has increased the amount of available information;
- Syntactic information can be analyzed by using Shannon's methods;
- Semantic information asks for meaning;
- The change of the indefiniteness can be used to evaluate information.

3.5 Complex Networks

Simple systems consist of only one level on which their agents are interconnected and interact. Such a horizontal scheme however is not sufficient to classify and order indefinite elements within a wider context. In order to represent the interactions of the agents, it is useful on the other hand to consider a *network* of interconnections. In this and the following section, I will therefore deal with systems which have a high degree of organization and which become complex through formation of a network. We are concerned here with the following problems: What types of networks occur? What defines the indefiniteness of a network? How does one locate an element within a network? Here, I will use the example of finding a search topic on the Internet, which represents the currently best-documented system of knowledge.

On the Web, the concept of vagueness as employed by language analysts is not associated with the question of "more or less". "Anton is bald" is a vague sentence because Anton's head has more hair than Benno's, who is completely bald, but less than Caspar's head, on which the few remaining hairs are well distributed. Although Williamson[50] and other analytic philosophers exclude quantitatively determinable concepts from the category of vagueness, the latter is situated on a linear scale, e.g. poor–rich, cold–warm, etc. Since this is too great a limitation to allow us to fix the meaning of a concept that we are seeking, the environment and the context of the concept have to be taken into account. The concept has to be captured within a network of interrelations. We shall therefore start with some general considerations about networks.

Carl von Linné (Linneaus) developed an example of a network in 1753, with his taxonomy of nature, which includes three "kingdoms": animals, plants and minerals. In the "kingdom plantae"[51] which was developed according to his plan, plants are classified into phyla, classes, series, then families, genera and finally species. The species "wild garlic" (or "bear's garlic"), for example, with the scientific name

[50] Williamson [42].

[51] Sterne and von Enderes [43]. For an English introduction to this topic, see e.g. Comstock [44].

"*allium ursinum L.*" according to Linné, belongs firstly to the genus of the onions ("*allium*"), then to the family of the lilies ("*liliaceae*"), which in turn belongs to the series "*liliales*" in the class of the monocotyledons ("*liliopsida*"), which itself belongs to the phylum of the flowering plants ("magnoliophyta"). This classification corresponds to a branched structure or tree diagram, as can be found also in political, legal or commercial organization charts. In order to classify an unknown plant XY, one has to answer a series of questions whose answers finally lead to the identification of the species. The tree diagram has a complex structure containing many branches and forks.

Allowing for certain limitations, every network can be represented by a graph, in which the nodes symbolize the agents and the edge lines symbolize the interconnections. In the above example, the nodes are the various classes, families, genera, and species, and the edge lines are the logical interconnections of the taxonomy. Mathematics has dealt with the theory of graphs and has developed a terminology with which to describe them. Physics investigates large networks empirically, and this has become much simpler in the past 10 years due to digital documentation using computers. A few basic concepts will be necessary to provide us with some insights into this fascinating field.

The nodes of the network are numbered consecutively. For each interconnection, a number (element) is entered into a column and a line of a rectangular array (matrix) corresponding to the nodes that it connects. A matrix element with the value 1 defines an interconnection between two nodes, while a matrix element with the value 0 means that no connection exists. As a reference for the structure of organized networks, a *random network* is employed, which is generated when the interconnections are chosen randomly.

Erdős and Rényi[52] have investigated the random networks which are obtained when each interconnection is made with the probability p. In their theory, the linking depends on this probability. Conversely, for a given graph with N nodes, out of the maximum possible number $N(N-1)/2$, and the actual number n of interconnections, the probability p can be computed. A node has the degree k if it has k interconnections. The probability that a node has k interconnections obeys a Poisson distribution with $\langle k \rangle = pN$, in the limit of large graphs. A node with many interconnections is called a *star* or a hub. Such stars can be seen on maps of airline routes, which distinguish certain cities (hubs) as important air traffic nodes. The cluster coefficient $C \approx p$ gives the probability that two neighbors of a particular node are themselves interconnected, i.e. that they form a triangle. The diameter of a random graph increases as the logarithm of the number of nodes.

Employing computers extensively, it has become possible to compare real networks with random networks. Albert-László Barabási[53] found drastic differences between real and random networks. Mark Newman's investigation of the publications by authors from a particular field showed that real networks have a much higher degree of organization than random networks. In this case, the network is

[52] Erdős and Rényi [45].

[53] Barabási [46].

defined by interconnecting all the authors of a given publication and including all publications from a particular field.

Newman investigated 66,652 publications by 56,627 high-energy physicists over a five-year period. He found a relatively large number, $\langle k \rangle = 173$, of co-authors. The empirically-determined number of co-authors, $\langle k \rangle = 173$, can be explained by the large experimental collaborations at high-energy accelerators, which often have as many as 1000 collaborators and which make a dominant contribution to this average value. The average number of co-authors of theoretical publications, in contrast, is only $\langle k \rangle = 2$. The average distance of two nodes is four, i.e. each author is connected to every other author via on average three mutual authors. The cluster coefficient of $C = 0.7$ deviates strongly from the value $C \approx p = 0.003 = \langle k \rangle / N$ predicted by the random theory.

The scientific world is "small", i.e. its authors are closely connected to one another. It was found in sociometric experiments in the USA that two randomly-chosen persons A and B are connected to each other through only six mutual acquaintances. The social psychologist Stanley Milgram,[54,55] has empirically investigated how persons (A) in the Midwest of the US contact a person (B) in Massachusetts, not known to them. They had to build up a chain of acquaintances. The average number of intermediate persons was five, and it varied between two and ten. On the average, such a chain had the following form: Person A knows person A1, he/she knows person A2 etc. (A-A1-A2-A3-A4-A5-B). We have all had the experience in party conversations of discovering that our guests know each other through diverse connections. Real networks exhibit this kind of 'neighborhood' of interconnections much more frequently than random graphs do.

Barabási and Albert formulated rules about how networks must be structured in order to have this property.[56] In each growth step, one node is added to the network. The new node is preferentially connected to a node that is already present and which has a large number of connections. This rule follows the "St. Matthew principle", which asserts that, "For he who hath, to him shall be given". The above authors derived a power law in their model for the distribution of the degrees of connection, $P(k) \propto 1/k^3$. We have already encountered power laws for the distribution of wealth. Since power laws lead to distributions that become monotonically smaller and have no maximum which would define their scale, such networks are also referred to as scale-free.

Now that we have gained some small insight into the structures of networks, we turn to the second question, that of the *indefiniteness* of a network.

Let us assume that we are dealing with a simple network consisting of 5 nodes. The nodes 1–4 form a square, to whose 4th corner the 5th node is attached. The connection matrix is a 5×5 matrix and its elements are zero where there is no connection and g when a connection is realized with the weight g. If one asks which degree of indefiniteness we associate with this graph, we can make use of the

---

[54]Newman [47].

[55]Milgram [48].

[56]Cf. Barabási and Albert [49].

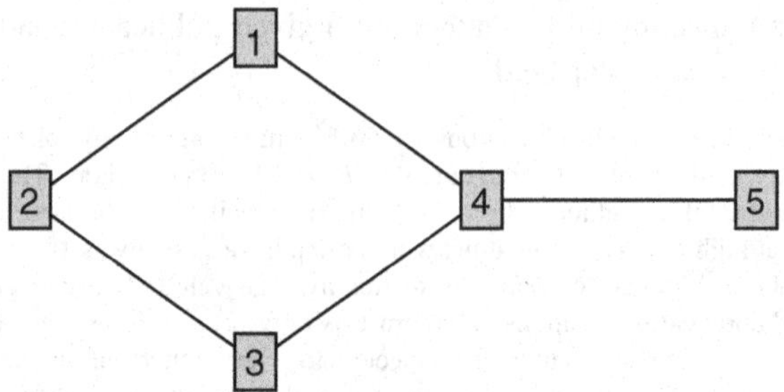

Fig. 3.5 A simple network with 5 nodes and 5 interconnections having the weights g. The weighting function $0 < g < 1$ for each interconnection shows the extent to which it is in fact realized. It plays the same role as the membership function in the case of linear indefiniteness (cf. Sect. 2.3)

definition of indefiniteness from Sect. 2.3 and generalize it to the connection matrix g[57]:

$$U(\text{graph}) = -\text{Trace}\,\{g\log[2, g] + (E - g)\log[2, (E - g)]\}\,.$$

As a simplification, Fig. 3.5 shows each connection with the same weight, $g = 0.5$; then the value of the indefiniteness is found to be $U = 10$, which corresponds to the 5 connections with this particular weight.[58] If we had chosen a more complex network such as the one in Fig. 3.6, where the fifth element enters into connections with four different nodes ($g = 0.5$), then the indefiniteness would be larger, namely $U = 16$.

In the graph in Fig. 3.6, the new element (5) has four indefinite connections to the four given elements of the original square, and the resulting indefiniteness is greater. It counts the number of indefinite possibilities with which each element is connected, in each case with the weight g = 0.5. The advantage of the general definition using the connection matrix g is of course that it can also give a well-defined answer for non-trivial weighting functions and networks. The answer in such cases cannot be estimated simply by counting.

What determines the form of a network? If one analyzes the elements and interactions of the empirical system, one can find a possible form for the network. But what are the criteria for such an analysis? In a cell, one can combine the molecules into a network according to their spatial nearness or their functionalities. On simple physical grounds, one would expect that nearby components also more strongly influence each other functionally. I know of no simultaneous analysis of these two possible networks (spatial, functional). Using two sets of weighting

[57] E defines a matrix which consists of only 'ones'; and the *Trace* Tr{…} is defined as the sum of the diagonal elements of the matrix given by the formula {…}.

[58] Log[2, 0.5] = −1.

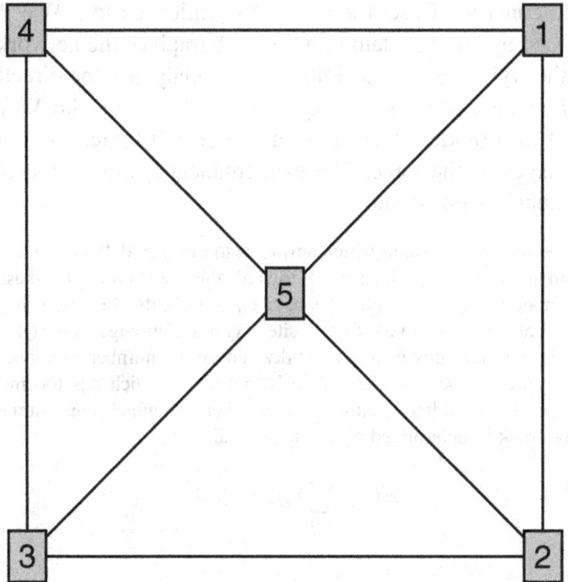

Fig. 3.6 A simple network with 5 nodes and 8 interconnections with weights g. If all the weighting functions have the value $g = 0.5$, the indefiniteness U is found to be 16

functions, such an analysis would yield two connection matrices whose similarity could then be investigated using the methods of mutual information (see Sect. 3.1).

It is presumed that the complexity increases with repetitions of motives of similar form in a network. The analysis of the eigenvalues of the connection matrix can classify different structure types.[59] In the case of random matrices, the eigenvalues are distributed in a semicircle around zero (Wigner distribution). Scale-free and hierarchical models have a strongly enhanced distribution at small eigenvalues. This is for example the case for the eigenvalue distribution of the connection matrix of a protein. It would be interesting to carry out an analysis of the indefiniteness as defined above for various empirical networks.

How does one locate an element within a network? Here, the indefiniteness is different from the previous case, where we computed the indefiniteness associated with the possible positions of a new node. For the present question, the node is already present in the network and we want to define a method of searching for it. Search engines attach *directions* to the interconnections, so that nodes with many connections can take on different significances. Many connections which point towards a node identify it as an "authority", while many connections pointing away identify a node as a transfer point or "hub" (hubs are known for example as central transfer points in air travel). The distinction between authorities and hubs helps to localize a search term in the World Wide Web.

[59]Cf. de Aguiar and Bar-Yam [50].

The principle consists of ascribing a high significance to a WWW search term which is referenced by an important hub. In a subgraph of the network,[60] each node is given an "authority" value and a "hubness" by analyzing the directed connection matrix $g(i \rightarrow j)$, which is no longer symmetric. A site on the Web with a high authority is used to provide information on the search topic, while a hub site can indicate other usages of the topic. The mathematical identification of the "authorities" and the "hubs" is not trivial.

Google therefore uses an algorithm which attributes to every node i a *page rank* $r(i)$. Every connection from another node j which points towards the site i is interpreted as a *vote* which is given by the node j. But not all the votes have equal weights; the voters are weighted by their own page ranks $r(j)$, i.e. a vote from a site j with a high page rank $r(j)$ counts with a higher weight. In addition, this weighting is divided by the number of connections which point away from the site, $k(j)$; the idea here is that a site which has too many outgoing connections is not such a reliable authority as another site which concentrates on certain topics. The page rank is determined by an iterative algorithm,

$$r(i) = \sum_{j \rightarrow i} r(j)/k(j).$$

It is nearly impossible to discover the criteria of the search engines for the Web. The Web itself gives only confusing indications as to how it functions. Without the help of introductory textbooks which summarize the information,[61] one is helpless. Since one cannot reconstruct these search algorithms oneself, it is wise to regard their results with some skepticism. Lee Smolin has said, "We used to cultivate thought, now we have become hunter gatherers of images and information. This speeds things up a lot but it doesn't replace the hard work in the laboratory or notebook which prepares the mind for a flash of insight".[62] In the same collection of answers to the question, "How has the Internet changed the way you think?", Kevin Kelly answered, "The Internet's extreme hyperlinking highlights those anti-facts as brightly as the facts. [...] My certainty about *anything* has decreased. Rather than importing authority, I am reduced to creating my own certainty—not just about things I care about—but about anything I touch, including areas about which I can't possibly have any direct knowledge. That means that in general I assume more and more that what I know is wrong".[63] Frank Schirrmacher has enriched the Internet discussion with suggestions about how we might win back control over our thinking: "We love unambiguousness, for the stronger it is, the stronger our feeling that we are in control. That is our way of dealing with risks. We develop routines which are similar to those of a computer. In our ambiguous surroundings, we have to evolve new categories;

[60]Manning et al. [51].

[61]For example: Manning et al. [51]; or Caldarelli [52].

[62]Smolin [53].

[63]Kelly [54].

in an unambiguous world, we would be prisoners of the categories. People have a need for things to be just so, and not otherwise".[64] If we identify indefiniteness with ambiguity, we are on the trail of the motivation of this study. Indefiniteness is the counterweight to information. Paying attention to indefiniteness improves the practice of science.

- An indefinite element can be localized as a node in a network;
- Real networks have denser neighborhoods than random networks;
- Orientation is given by "authorities" and "hubs";
- Indefiniteness as an antidote to the apparent unambiguity of the Internet.

3.6 The Neural Network

A neural network is a network which is modeled on the brain. As such, it models the action potentials which are exchanged between nerve cells. These potentials, which lie in the millivolt range, can be measured using microprobes. Thus, the functioning of a neural network can be verified for simple processes. In general, a "neural network" can be a computer model which is used for data analysis and for making predictions. It "learns" through input data and the resulting outputs. When the learning phase is completed, the network generates relatively reliable predictions for new input data which have not yet been processed. In the following, first I will sketch the model of a single neuron in the brain, the *perceptron*, and then the bundling of a number of such units into a large network. The question of how well such a network can simulate general, complex problems will be discussed.

For a discussion of the manifold processes in the brain, the reader is referred to Sect. 2.5. The brain is extremely complex and contains some 10,000 million neurons which are fed with information via their dendrites and output them through their axons. Donald Olding Hebb[65] developed a computer model of a single neuron, the *perceptron*. The program written by Hebb contains parameters ("weights") which are optimized with respect to the input data. Thereafter, the program uses these optimized parameters to make predictions. In the following, I will employ the vocabulary of artificial intelligence and will personalize the programmed perceptron. The perceptron "learns" to either amplify or attenuate input potentials $x(i)$ ("stimuli") by varying the weights $w(i)$. A large and positive weight amplifies the associated input potential, considering it as particularly important. A negative

[64] Schirrmacher [55].

[65] Hebb [56].

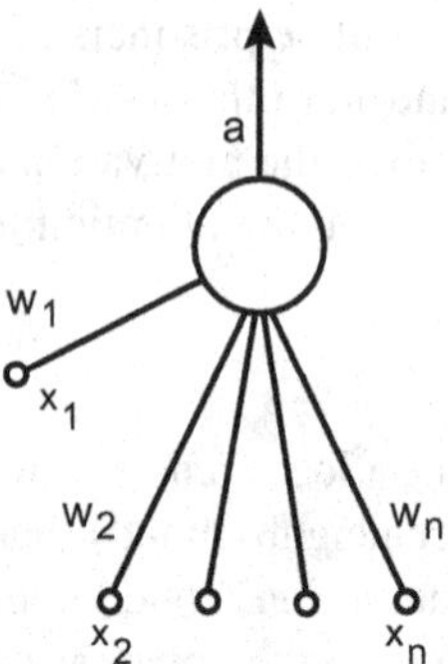

Fig. 3.7 A model of a perceptron with the input channels x_i, the weighting factors w_i, the activation parameter a and the output signal $y(a)$. The dendrites form the input channels, while the strengths of their synapses are modeled by the weighting factors; the axon conducts the output signal onwards (cf. also Sect. 2.1)

weight reduces the contribution of the corresponding potential to the output signal. The sum of such responses is the *activation* parameter a of the neuron:

$$a = w_1x_1 + w_2x_2 + w_3x_3 + w_4x_4 + \cdots .$$

The activation parameter a determines the non-linear activation function $y(a)$ ("reaction"), which varies between 0 and 1 depending on the strength of the activation parameter. The extreme values $y = 1$ or $y = 0$ mean that the neuron fires or does not fire. In principle, intermediate values are possible. The activation parameter carries the information which the neuron has received (Fig. 3.7).

In order to describe the functioning of such a network program, it is best to consider an example. How can a neural network answer the question as to whether an arbitrary 6-letter word belongs to the German language? An allowed (not allowed) word is associated with a target value $t = 1$ ($t = 0$). Based on 1000 non-existent and 1000 existing words, each with six letters[66] x_1, x_2 … x_6, the neural network learns to adjust the weights w_1,w_2 … w_6 in such a way that the deviation between the output function $0 < y(a) < 1$ and the target function t is minimized. In the ideal case, the algorithm finds the optimal weights. Using these weights, the neural network can then treat and classify new words. David J.C. MacKay[67] shows the adjustment of the weights after 30, 300, and 3000 iterations. For two input data, two weights are adjusted; they define a straight line (see Fig. 3.8, left side). This line separates the positive cases (+) from the incorrect cases (0). The procedure converges rapidly for a modest number of weights. However, it may happen that the form of the parameters permits no solution (Fig. 3.8, right side). The problem is then not separable and one has to choose different parameters in order to find a solution. This is similar to a change in perspective, without which often no progress

[66]The six letters of the word are identified by six numbers, e.g. A = 0, B = 1, …, Z = 26.
[67]MacKay [57].

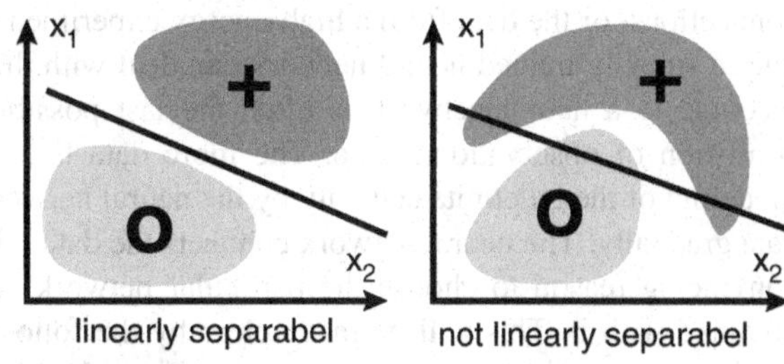

Fig. 3.8 Examples of linearly separable and non-separable problems with two input channels, x_1 and x_2

can be made in a cognitive process. The above example of the identification of an unknown word makes the difference between a neural network and a search algorithm clear. With a search algorithm, the computer is programmed to search for the input word in its memory using an input address. If it is found, the word is allowed. The neural network, in contrast, learns in general how German words are constituted and computes its result based on the abstracted weights. The procedure used by the neural network is termed "intelligent", because it can, through a new learning process for example with Italian words, then also decide whether a word belongs to the Italian language. A "dumb" search routine with a German dictionary stored in its memory would not be able to achieve such a result.

Modern neural networks are tools which can evaluate data in a non-linear way. They model the complex relationships between input and output data in order to recognize patterns. More complicated networks contain units composed of several basic building blocks which bundle the input data in one layer before they output a result. Different network layers can then be connected in series. It is even possible to determine the topology of the network itself, if the number of network nodes is not too large.

In a simplified version of the neural network, the firing rates are divided into two categories: "yes" means that the neuron fires, and "no" means that it does not fire. John Joseph Hopfield[68] has constructed an analogous spin system, for which he defines an energy functional. The strengths of the synapses produce a potential landscape. There is no single minimum in which the system preferentially comes to rest, but there are several almost equally preferred metastable states which permit the system to become an effective pattern-storage device, so that it functions like a *memory*. When a learned pattern is found to correspond to an external stimulus, then the neural system is drawn towards this ideal state (attractor), i.e. the memory focuses on the corresponding content.

In mathematical modeling of neural networks, (secondary) states occur which correspond to indefinite patterns. Learning produces its own confusion. Experiments have shown that similar secondary states can exist in the brain. Sleep plays an important role in managing these indefinite patterns; it suppresses their influence and thereby improves the performance of memory.

Attempts have been made using neural networks to simulate systems which are too complex to understand on a microscopic level. The organization of intercellular

[68]Hopfield [58].

connections, or the data from a high-energy experiment, can be so complicated that only a suitably trained neural network can deal with them. Along with fuzzy logic, recourse to a neural network is often the last possibility to bring order into the confusion of observational data. The more data that are available, the better the modeling of the indefinite actuality by the neural network. It approaches the definite data gradually. The neural network connects the data with the object, but there is no convincing reason to choose the particular network and thus to favor the results obtained with it. This will be made clear by the following example.

If one gives a neural network the starting velocities and angles of a golf ball, together with the resulting landing points, as input data for learning, then after several iterations it can compute the corresponding trajectory for two new initial values. Although the weighting functions and the "wiring" of the neurons are known, the neural network still remains a "black box" which knows nothing about the gravitational force or the laws of motion. It would be fantastic if a computer could develop the ability to make abstractions—and the creativity—of a Newton or Galileo. However, a generalized understanding of the underlying causes of the trajectory is not accessible to a neural network. A neural network cannot help us to understand gravitation. Defenders of the computer approach may object that the physical description is only a simplified model, without frictional forces or wind velocity, while complications of this type can be input automatically with the data into the network and thus are taken into account. The attempt has been made to predict stock market prices using neural networks, but this was a failure. However, in this respect, artificial intelligence is no worse off than human intelligence.

The wide compass of the last six sections has stretched from the most primitive units of information theory, the bits, up to the artful and elaborate construction of neural networks. The grandiose evolution of modern information technologies was summarized and it was shown that by making use of the indefinite, a qualitative evaluation of information is attainable. I chose the title "*Approaching* the definite" for the present chapter because the methods applied here bind the indefinite only weakly to the definite. A neural network learns on the basis of an abstract program architecture whose switches can be adjusted appropriately; explicit weights for the couplings were not formulated. This variability is both an advantage and a disadvantage. Programmed learning seeks its own paths to optimization, but for many purposes, this occurs in an overly uncontrolled manner. We will therefore now turn to stricter methods of analysis and comprehension which consider the indefinite together with the definite.

- A neural network interconnects neurons on different levels;
- A perceptron is a model of a neuron;
- The learning process consists of adjusting the strengths of synapses ("weights");
- A neural network connects data and object within a "black box".

References

1. Shannon, C.E.: A mathematical theory of communication. Bell Syst. Tech. J. **27**, 379–423, 623–656 (1948)
2. MacKay, D.J.C.: Information Theory, Inference and Learning Algorithms, p. 25. Cambridge university press, Cambridge (2003)
3. Hagenauer, J.: Texte aus der Sicht der Informationstheorie – Wie viel Information enthalten Texte und wie kann man Autoren ermitteln? In: Bayerische Akademie der Wissenschaften, vol. 02, pp. 50 ff (2007)
4. Andersen, C.: The end of theory: the data deluge makes the scientific method obsolete. Wired Mag. **16**(07) (2008)
5. Bronstein, I.N. et al.: Taschenbuch der Mathematik, Stuttgart-Leipzig 1996, p. 1098. English: Abramowitz, M., Stegun, I.A.: Handbook of Mathematical Functions. http://people.math.sfu.ca/~cbm/aands/abramowitz_and_stegun.pdf, p. 931
6. Grupen, C.: Astroparticle Physics, p. 403. Springer, Heidelberg (2005)
7. Wootters, WK., Zurek, W.H.: A single quantum cannot be cloned. Nature **299**, 802 (1982). See also: http://de.wikipedia.org/wiki/No-Cloning-Theorem
8. Kac, M.: Can one hear the shape of a drum? Amer. Math. Monthly, Part 2: Papers Anal. **73**(4), 1–23 (1966)
9. Gordon, C., Webb, D.L., Wolpert, S.: One cannot hear the shape of a drum. Bull. Amer. Math. Soc. **27**, 134–138 (1992)
10. Wheeler, J.A.: Information, physics, quantum: the search for links, In: Zurek, W.H. (ed.) Complexity, Entropy, and the Physics of Information. California (1990)
11. Unger, P.K.: The Problem of the Many. Midwest Stud. Philos. **5**, 411–467 (1980)
12. Wittgenstein, L.: Tractatus Logico-Philosophicus, Satz I. Frankfurt am Main 1963. English: see http://people.umass.edu/klement/tlp/
13. Zeilinger, A.: Einsteins Schleier – Die neue Welt der Quantenphysik (Einstein's Veil. The New World of Quantum Physics), p. 231. München (2003)
14. Chin, A.W., Dotta, A., Caruso, F., Huelga, S.F, Plenio, M.B.: Noise-assisted energy transfer in quantum networks and light-harvesting complexes (2010). arXiv:0910.4153[quant-ph]
15. Everett, H.: Relative state formulation of quantum mechanics. Rev. Mod. Phys. **29**, 454–462 (1957)
16. Susskind, L.: The Cosmic Landscape: String Theory and the Illusion of Intelligent Design, New York (2003)
17. Aguirre, A., Tegmark, M., Layzer, D.: Born in an Infinite Universe: a Cosmological Interpretation of Quantum Mechanics (2010). arXiv:1008.1066[quant-phys]
18. Lewis, D.K.: On the Plurality of Worlds. Oxford, New York (1986)
19. Descartes, R.: Aus den Regeln zur Leitung des Geistes (Original title: Regulae ad Directionem Ingenii), Frankfurt am Main, Rule, vol. 4, p. 32 (1960). English: see http://en.wikipedia.org/wiki/Rules_for_the_Direction_of_the_Mind
20. Luhmann, N.: Theorie der Gesellschaft oder Sozialtechnologie. Frankfurt am Main, p. 295 (1971) (With Habermas, J.; not available in English.)
21. Li, M., Vitányi, P.: An Introduction to Kolmogorov Complexity and its Applications. New York (1993)
22. Bennett, C.H.: Dissipation, information, computational complexity and the definition of organization. In: Pines, D. (ed.) Emerging Syntheses in Science, pp. 297–313. Santa Fe Institute (1985); Reading, Massachusetts (1987)
23. Lopez-Ruiz, R., Mancini, H.L., Calbet, X.: A statistical measure of complexity. Phys. Lett. A **209**, 321–326 (1995)
24. Odum, E.P.: Ecology, p. 29, New York (1963)
25. Mainzer, K.: Thinking in Complexity: The Computational Dynamics of Matter. Mind and Mankind, Berlin (2007)

26. Dürr, H.P., Zimmer, S. (camera): Wir erleben mehr als wir begreifen – Naturwissenschaftliche Erkenntnis und Erleben der Wirklichkeit. Clausthal-Zellerfeld (2002). http://video.tu-clausthal.de/vortraege/duerr2002/. See also http://www.bibliotecapleyades.net/ciencia/ciencia_fisica45.htm, and http://en.wikiquote.org/wiki/Max_Planck
27. Goldenfeld, N., Woese, C.: Biology's next revolution. Nature **445**, 369 (2007)
28. Parisi, G.: On Complexity (unpublished); see also: Complex Systems: a Physicist's Viewpoint (2002), arXiv:cond-mat/0205297v1 [cond-mat.stat-mech]
29. Mitchell, S.: Komplexitäten. Warum wir erst anfangen die Welt zu verstehen. Frankfurt am Main (2008)
30. Mitchell, S.: Unsimple Truths: Science, Complexity and Policy. University of Chicago Press, Chicago (2009)
31. Luhmann, N.: Einführung in die Systemtheorie. In: Baecker, D. (ed.) Lectures 1991/92, pp. 67 ff. Heidelberg (2009)
32. Luhmann, N.: Die Wissenschaft der Gesellschaft, Frankfurt am Main (1990). English translation of chapter 10: The modernity of science. New German Critique **61**, 9–23 (1994)
33. Moore, G.E.: Cramming more components onto integrated circuits. Electronics **38**(8), 114–117 (1965)
34. Berners-Lee, T.: Information management: a proposal (1990). http://www.w3.org/History/1989/proposal.html
35. Friedell, E.: Kulturgeschichte der Neuzeit, p. 661. München (1960). English: A Cultural History of the Modern Age. New Brunswick, NJ (2008)
36. Putnam, H.: The meaning of 'meaning' Mind. In: Language and Reality, Philosophical Papers, vol. 2, p. 215. Cambridge (1975)
37. Miller, F.J.: I = 0 (Information has no intrinsic meaning). Inf. Res. **8**(1), no. 140 (2002)
38. MacKay, D.M.: Information Mechanism and Meaning, p. 85. The MIT Press, Cambridge (1969)
39. Pariser, E.: The Filter Bubble: What the Internet is Hiding from You, New York (2011)
40. Luhmann, N.: Einführung in die Systemtheorie. In: Baecker, D. (ed.) Lectures 1991/92, p. 129. Heidelberg 2009
41. Heidegger, M.: Der Satz vom Grund. In: Jaeger, P. (ed.) Heidegger Complete edition, vol.10, p. 189. Frankfurt am Main (1997). English: Heidegger, M.: The Principle of Reason. Indiana University Press, Bloomington (1996)
42. Williamson, T.: Vagueness, London and New York (1994)
43. Sterne, C., von Enderes, A.: Unsere Pflanzenwelt. Blumen, Gräser, Bäume und Sträucher, Pilze, Moose und Farne der mitteleuropäischen Flora, p. 532. Berlin (1955)
44. Comstock, J.L.: An Introduction to the Study of Botany: Including a Treatise on Vegetable Physiology, and Descriptions of the Most Common Plants in the Middle and Northern States. Robinson, Pratt & Co., New York (1837)
45. Erdős, P., Rényi, A.: On the evolution of random graphs. Pub. Math. Inst. Hung. Acad. Sci. **5**, 17–61 (1960)
46. Barabási, A.L.: Linked—How Everything is Connected to Everything Else and What It Means for Business, Science, and Everyday Life, New York (2003)
47. Newman, M.E.J.: Scientific collaboration networks: I. Network construction and fundamental results. Phys. Rev. E **64**, 016131 (2001)
48. Milgram, S.: The small world problem. Psychol Today **1**(1), 61–67 (1967)
49. Barabási, A.L., Albert, R.: Emergence of scaling in random networks. Science **286**, 509 (1999)
50. de Aguiar, M.A.M., Bar-Yam, Y.: Spectral analysis and the dynamic response of complex networks. Phys. Rev. E **71**, 016106 (2005)
51. Manning, C.D., Raghavan, P., Schütze, H.: Introduction to Information Retrieval. Cambridge University Press, Cambridge (2008)
52. Caldarelli, G.: Scale-Free Networks: Complex Webs in Nature and Technology. Oxford (2007)

53. Smolin, L: We have become hunters and gatherers of images and information. In: Annual Question, How Has The Internet Changed The Way You Think? (2010). http://www.edge.org/q2010/q10_4.html#smolin
54. Kelly, K.: An intermedia with 2 billion screens peering into it. In: Annual Question (2010). http://www.edge.org/q2010/q10_1.html#kelly
55. Schirrmacher, F.: Payback, p. 179. Munich (2009)
56. Hebb, D.O.: The Organization of Behavior: A Neuropsychological Theory. New York (1949)
57. MacKay, D.J.C.: Information Theory, Inference and Learning Algorithms, p. 471 ff. Cambridge University Press, Cambridge (2003)
58. Hopfield, J.J.: Neural Networks and Physical Systems with Emergent Collective Computational Properties. Proc. Nat. Acad. Sci. USA **79**, 2554 (1982)

Chapter 4
Establishing the Definite from the Indefinite

Human nature inclines us to ignore indefiniteness, to underestimate it, or to hide it behind explanations, theories and absurdly small probabilities. When Richard P. Feynman[1] was preparing an assessment of the Challenger space shuttle accident of 1986, he asked engineers and managers from NASA, "How high is the probability of such an accident occurring?" While the engineers would have expected a probability of around 1/100, the managers estimated it to be 1/100,000—a value that is obviously much too small. This absurdly small number would mean that a manned rocket could be launched into space once every day for 300 years without an accident ever occurring.

This example, together with many others, teaches us to treat estimated probabilities with caution. Such probabilities are imprecise and must be carefully scrutinized. It is my intention in this chapter to consider indeterminacy as the explicit starting point for analysis and decision-making. I will discuss various methods for achieving this: empiricism, decision theory and fuzzy logic are 'hard' physical, mathematical and logical methods. Metaphors, sign theory and scenarios are 'soft' hermeneutic concepts. Sign theory plays a particularly important role here, since it interprets the indefinite object in terms of its definite data. It demonstrates the theme of this essay in a highly concise manner, namely that the definite and the indefinite should be considered together.

4.1 Experimental Methods

Which obstacles arise when we attempt to determine objects and phenomena? To the extent that new empirical results push back the limits of our knowledge, tension will occur between what is considered to be established knowledge and the new insights. Seen in a general context, the older constructions may be called into question and the determinate may appear to be slipping away, like the image of a black vase used in gestalt psychology, which switches over to an image of two

[1] Richard Phillips Feynman: *Feynman's Appendix to the Rogers Commission Report on the Space Shuttle Challenger Accident*. http://www.ralentz.com/old/space/feynman-report.html.

H.J. Pirner, *The Unknown as an Engine for Science*,
The Frontiers Collection, DOI 10.1007/978-3-319-18509-5_4

white faces while we are looking at it. In this section, I will trace out the role of experiments and measurements, which have often given rise to such upheavals.

Hans-Jörg Rheinberger holds that the significance of experiments lies in their concentration of our attention onto a particular question. This self-limitation makes it possible to perceive what is still unexplored. "New knowledge is generated not so much in the minds of the scientists—where indeed it must in the end come to fruition—but rather in the experimental system itself [...]. Experimental systems are thus extremely subtle arrangements; but one must consider them to be points of emergence which we have designed to capture what is not completely imaginable. They are like spiders' webs. It must be possible for something to become trapped in them, about which we lack exact knowledge of what it is and when it will occur. François Jacob [spoke of them as] 'machines for generating the future'".[2]

Ernst Cassirer, in his *Philosophy of Symbolic Forms*, describes the determination of scientific concepts as a "representation which is contained in an infinite set of different possible concepts in terms of their common characteristics. If those characteristics are to be [...] determined, [...] it would seem that the most certain procedure [...] is in fact to parse through the set whose common characteristics are being sought. One may simply place the elements one beside the other, and in the sheer act of enumerating them will recognize the form of their common unity, that which binds them together".[3] In my opinion, it is the series of *indeterminate* elements which most effectively contributes to the development of new research results. The experimental system is the generating machine that produces these elements. The measurement results are the set within which the special cases are united. Undefined special cases are especially important at the beginning of every new scientific discovery. An example is the discovery of the superfluid phases of ^{3}He: In the course of his doctoral work, Douglas Dean Osheroff[4] observed phenomena at very low temperatures which were completely unexpected for ^{3}He atoms. He confirmed these phenomena in further experiments and was able to identify them as arising from a condensate consisting of pairs of ^{3}He atoms. His discovery was rewarded with the Nobel Prize in 1996.

A measurement yields neither a large nor a small number; that would require fixing a scale to allow the comparison of various results. It is important to distinguish measurements from vague determinations. At the end of the 19th century, physicists and chemists were confronted with the problem of atomic weights. The increasing precision of their measurements revealed limiting cases which seemed not to fit into the scheme of atomic theory. In the Periodic Table, the atomic weight[5] of platinum is given as 195.09 atomic mass units, relative to the ^{12}C carbon atom. The rounded integer number 195 would indicate the number of nucleons (protons and neutrons) within the platinum nucleus, just as there are 12 nucleons in the

[2]Rheinberger [1].

[3]Cassirer [2].

[4]Osheroff [3].

[5]Barber [4].

carbon nucleus. In particular, the question arose as to why the atomic weight of oxygen, 15.9994, is so close to the integer 16, while that of chlorine, 35.453, is far from being an integer number. Only after the development of nuclear physics did it become clear through investigations of radioactivity in heavy elements that chemically identical substances may differ in their nuclear structures and thus may have different atomic weights.

Nature permits the existence of different atomic nuclei with the same electric charge, i.e. the same number of protons, but differing numbers of electrically uncharged neutrons. These nuclear twins which have the same number of protons are called *isotopes*, and they exhibit identical chemical behavior. The unusual atomic weight of 195.09 for platinum indicates that when we weigh platinum atoms, we are not always dealing with nuclei containing 195 nucleons, i.e. 78 protons and 117 neutrons. Platinum occurs in nature as a mixture of isotopes, and this is the solution to the riddle. The hypothesis that a still unknown coordinate or a not-yet-understood parameter exists is typical of scientific thinking when it is attempting to deal with undefined limiting cases. Neutrons were first produced by Walther Bothe and Herbert Becker in 1930 by bombarding beryllium with alpha particles (and first identified in the scattering experiments of James Chadwick in 1932). Their discovery raised the question as to what force binds the nucleons together in the atomic nucleus. Only five years later was the hypothesis proposed by Hideki Yukawa that the nuclear force results from the exchange of mesons (pions).

> Historically, one can discern two different paths along which the modern natural sciences have evolved. On the one hand, there is the purposeful, slow, day-to-day work of many thousands of scientists, who painstakingly accumulate experiments and theories, in order to construct a new edifice of knowledge piece by piece. On the other hand, there are the revolutionaries, who find a new paradigm at a single stroke, and thus solve problems which have persisted for decades or even centuries. In reality, anomalies and indefiniteness often play an important role. Unusual phenomena are not sufficient to falsify a theory. However, experimental results can lead a theory into difficulties by showing how limited its predictive power is. Nuclear physics offers a good example of this; although it has been pursued intensively since the work of Hideki Yukawa in 1935, the binding of nucleons into a nucleus is still difficult to explain within the Standard model of elementary-particle physics.

Physicists understand the theory of the strong interaction that binds the quarks and holds the proton and the neutron together; but they have difficulties in describing atomic nuclei, which can be up to ten times larger than individual nucleons.

If the determination of an object is tentative, its specific content remains flexible and it can be adjusted to fit new facts as they become available. This leads to a “positivity of the indeterminate”,[6] as derived by Gerhard Gamm from Hegel’s dialectics. Gamm sees something positive in the inscrutable nature of the modern world, since it admonishes us to be more cautious and limits our potential for amoral behavior. Michel Serres suggests a fluid approach to the theoretical development of efforts to increase knowledge: “Indeed, nothing is more difficult to

---

[6]Cf. Gamm [5].

imagine than a free and fluctuating time which is not yet completely determined, in which researchers basically do not know what they are searching for, although they unconsciously in fact do know".[7]

From the position of the natural sciences, it is difficult to understand this post-modern point of view. I would rather ask the reverse: Are there instruments within the humanities and the social sciences which correspond to the experimental machinery of the natural sciences? Surely, those fields have their own methods for arriving systematically at new knowledge. Empirical investigations in sociology and political science have similar goals to experiments in physics. The constructed situation makes it possible for psychologists or sociologists to see the objects of their research in a new light. This indeed concentrates the attention of the researchers onto a limited domain; but they can never be absolutely certain that they can control all of the variables of the problem. The philosopher Ernst Cassirer sees "…this absolute trust of the scientist in the reality of things" become "…shaken as soon as the problem of truth in the humanities enters upon the scene". Is there a theoretical method for finding the truth? The truth is definite. How must we proceed when our starting point is indefinite? If we cannot find truth, how can we endeavor to secure truthfulness? Here, epistemological and operational-theoretical questions merge together.

Ernst Cassirer raises the objection that a simply set-theoretical definition is insufficient; rather, a relational or functional connection must be discovered. "Thus, the attempt to make the content of a concept understandable by starting from its extension can no longer be maintained".[8] The indeterminate objects cannot simply be classified within a set of attributes, but rather must be fitted into the context of other objects which surround them and which characterize them. We have identified this context as a network (see Sects. 3.5 and 3.6). Cassirer describes the cognitive judgment in the Kantian tradition as an analytics of reason which intends to show "… how the various fundamental forms of cognition, such as sensory perceptions and pure intuition, such as the categories of pure reason and the ideas of pure reason, interlock with each other—and how they determine the theoretical form of reality in their interactions and their interactive designations. This *determination* is not simply captured from the object, but rather it includes in itself an act of the spontaneity of reason".[9] Nassim Nicholas Taleb expresses himself more critically: "…we are explanation-seeking animals who tend to think that everything has an identifiable cause and grab the most apparent one as *the* explanation".[10] As long as the determination is not substantiated with an underlying reason, it remains tentative. In Sect. 4.5, I will again pose the question of what causes the indefinite and the definite to unite.

---

[7]Authier [6].

[8]Cassirer [2, p. 343].

[9]Cassirer [2, p. 7].

[10]Taleb [7].

- Considering the definite and the indefinite together means: posing the question of what causes them to unite;
- The experimenter awakens the correlation between the definite and the indefinite;
- The relational reference, an act of spontaneity?

4.2 Decision Theory and Game Theory

The scientist who is striving for the highest-possible level of clarity has different goals from a practitioner in the functional world who must follow certain rules, laws or norms. Many problems with indefiniteness are related to decisions which must be taken in an uncertain, ambiguous or risky situation. Technical decisions are made with well-known input data and functional goals, so that their control can be left to machines. An individual, however, has to define his or her goals before taking action. One can investigate the norms or behavior involved in taking decisions; descriptive decision theory studies the psychological factors which influence the practical decisions of people in various economic, political or social contexts. In the following, however, we shall deal only with the normative criteria which permit rational decisions to be made starting from an indefinite situation.

As we have seen, sources of indefiniteness can be complex and multifarious. Too much information, uncertain data, contradictory indications or vague concepts may generate indefiniteness. Likewise, a decision may lead to several different goals which can even be mutually contradictory. In medicine, the physician may have to choose between a therapy which will extend the life of a patient at all costs, or one which will improve his quality of life, or one which will reduce his pain. In order to establish a goal function, the different goals are first ordered qualitatively according to priorities and then weighted by their importance.

For decision-making under uncertain conditions, there are uncontrollable factors which influence the outcome. Bad weather can destroy the harvest of a farmer, for example. Probabilities can help us to quantify those uncertainties. Let us assume that there are two alternatives a_1 and a_2, which as a result of uncontrollable factors can occur with the probabilities $p_1 = 0.8$ and $p_2 = 0.2$. The alternatives could for example be planting rye or wheat. The uncontrollable factor here is the weather, which will be dry with the probability p_1, or rainy with the probability p_2. The profit per 100 kg of the grain planted is shown in the matrix M, taking into account the various alternatives (see Table 4.1).

Table 4.1 The scheme for arriving at a decision contains the states corresponding to dry and wet weather, which occur with the probabilities p_1 and p_2, and the decision alternatives a_1 and a_2, planting rye or wheat

	Weather (dry), p_1	Weather (rainy), p_2
Alternative a_1 (rye)	$M_{11} = 10$	$M_{12} = 10$
Alternative a_2 (wheat)	$M_{21} = 15$	$M_{22} = 8$

The corresponding expected profits (€ per 100 kg) are shown as matrix elements

If rye is planted (alternative a_1), the profit in case of dry weather (with probability p_1) will be $M_{11} = 10$ €, and in case of wet weather, with the probability p_2, it will also be $M_{12} = 10$ €; corresponding values for the profits when wheat is planted (alternative a_2) are for dry weather 15 €, and for rainy weather 8 €. How should one decide? The average profit for the two alternatives is obtained from the sum of the profits weighted by their probabilities:

$$\langle \text{Profit for alternative } a_1 \rangle = M_{11} \times p_1 + M_{12} \times p_2 = 10\,€;$$
$$\langle \text{Profit for alternative } a_2 \rangle = M_{21} \times p_1 + M_{22} \times p_2 = 13.6\,€.$$

If we assume that the decision can be repeated an infinite number of times, we should choose alternative a_2, which offers a higher net profit. After a finite number of repetitions, the expected profit will be obtained. For the farmer, however, the fluctuations of the weather are important. In case of rain, with the second strategy he is risking a loss of 2 €/100 kg. A cautious decision-maker would therefore prefer alternative a_1, since then a profit of 10 €/100 kg is guaranteed.

These considerations can be improved by introducing a function which parameterizes the *gain* as a function of the risk. An increase in profits from an investment always represents a major gain for a small amount of net assets, while the same increase will approach a limiting gain when the net assets are much greater. The form of the risk-gain function depends on the risk attitude of the investor. The risk-gain function can be used to compute the expectation value of the net gain; it is a better indicator than the expectation value of the profit alone.

In the theoretical discussion, both the number of possible states as well as the probabilities of each state must be considered. In the above example, there might be extreme states in addition to dry or rainy weather, e.g. extreme drought or flooding. How well known are the probabilities for such extreme states? With lotteries or symmetrical games of chance, like tossing a coin, the probabilities are known in advance owing to the equivalence of all the possibilities. In the case of the weather, one has to try to take into account empirical data from previous years in order to estimate the probabilities for future events. Particularly difficult are decision processes involving extremely rare events which are associated with very high losses or very large gains. The expectation value for the loss due to such an event yields an undefined result obtained by taking the product of the probability "zero" with the loss value "infinity". An example is the "Worst Case Scenario" (WCS) for an accident involving a nuclear reactor, i.e. a reactor core meltdown. Such an event can occur as the result of a series of minor hazardous incidents which are treated as independent of each other in computing their probabilities. In fact, the various incidents are often correlated with each other.

An exacerbation of the decision-making situation is found when decisions must be made under conditions of incertitude, in contrast to ambiguity. The term *incertitude* is used when no initial probabilities are known at all, or when they are partially inaccessible. There are various rules for attacking this problem,[11] and one has to take the optimism or pessimism of the decision-maker into account. The decision-maker can consider his own feelings and determine his own subjective "probabilities". An advisor or consultant working in an ambiguous situation usually also takes the expectations of the decision-maker into account.

In economics, sociology and politics, problems must often be solved in which the result of an action depends not only upon the action itself and the situation, but also on the simultaneous or resulting actions of others. In this case, *game theory* has become established as a method for developing strategies to find the optimal course of action. These studies have been recognized by Nobel prizes[12] for economics. Game-theoretical concepts are also used in evolutionary biology.[13] In contrast to the decision problem sketched above, at least two players, 1 and 2, must be considered, whose gains are represented by two numbers, i.e. the gain of player 1 and the gain of player 2.

A very popular example is the "Prisoners' Dilemma": The police have arrested two suspects, whose roles in the crime are not clear. They are questioned individually and offered the following deal: Whoever testifies against the other prisoner will be released without a prison sentence (gain = 0), while the other prisoner will receive a prison term of 10 years (gain = −10) if he refuses to talk. If each prisoner testifies against the other, then each will receive a prison sentence of 5 years (gain = −5). If both refuse to talk, each will receive a prison term of 1 year (gain = −1). The prisoners are not allowed to speak to each other to agree upon a common strategy. The gain has negative values since a gain of (−10), corresponding to a 10-year prison term, is less than the gain of (−1) for 1 year of prison. The strategies followed by each prisoner are symbolized by "silent" and "testifies". The result for each prisoner is shown in a matrix where each element gives the gain for player 1 and for player 2, depending on their chosen strategies (cf. Table 4.2).

[11]Cf. Dörsam [8].

[12]Robert John Aumann and Thomas Crombie Schelling (2005) for the study of conflict and cooperation using game theory; John Forbes Nash, Reinhard Selten and John Charles Harsanyi (1994) for the role of chance in complex systems.

[13]Cf. Smith [9].

Table 4.2 The strategies of player 1 are shown vertically, while the strategies of player 2 are shown horizontally

	Player 2 *silent*	Player 2 *testifies*
Player 1 *silent*	−1, −1	−10, 0
Player 1 *testifies*	0, −10	−5, −5

The matrix elements give the pairs (gain for player 1, gain for player 2) corresponding to the combination of their two game strategies

In order to pursue his own self-interest, each player can attempt to maximize his own advantage independently of the behavior of the other player. Each would thus assume that the other player is always seeking the worst result for him. Prisoner (1) considers the consequences of remaining silent and sees that it could lead to the longest prison sentence (−10). If he pursues the second strategy and "testifies", then the worst case is a gain of (−5). He will thus decide in favor of the second strategy with a shorter prison term. Prisoner (2) will do the same. The result of this game, played purely rationally, is then mutual accusation with a 5-year prison sentence for each player (cf. Table 4.3).

Table 4.3 The value (gain for player 1, gain for player 2) = (−5, −5) for the "optimum" strategy of both players according to the criterion of maximizing gain independently of the strategy of the other player

	Player 2 *silent*	Player 2 *testifies*
Player 1 *silent*		
Player 1 *testifies*		−5, −5

Both prisoners receive 5-year prison terms

The above game represents an equilibrium strategy,[14] since neither player can improve his gain by one-sidedly abandoning the strategy. In this sense, equilibrium represents a solution to the game-theoretical problem.

The paradox results from the fact that in the above case, a clearly better strategy for both prisoners would be to remain silent, which would lead to just one year of prison for each. This strategy of course presumes that they have reached a prior agreement not to betray their cooperation by testifying against each other. In real economic life, it is possible to make such a cooperation binding by signing a contract, and thereby to take advantage of the cooperative strategy. The two prisoners however are not permitted to employ this possibility (cf. Table 4.4).

[14]Cf. Nash [10].

Table 4.4 The best strategy (gain for player 1, gain for player 2) = (−1; −1) for both players is for both prisoners to remain "silent"

	Player 2 *silent*	Player 2 *testifies*
Player 1 *silent*	−1, −1	
Player 1 *testifies*		

This strategy leads to only 1-year prison sentences for each; however, it presumes that there is an agreement between the players not to accuse each other

As normative disciplines, both decision theory and game theory are closely related to ethical questions, i.e. the gain must not always be in material form. Ecological goals or the public good can very well enter into the deliberations. From analytic philosophy, David K. Lewis asserts that language conventions represent equilibrium states in communication games.[15] Game theory has found numerous applications as a mathematical method of dealing with indefiniteness, in addition to fuzzy logic and the theory of neural networks. In game theory, priorities can be represented by natural numbers, which is often simpler than attempting to quantify membership functions on a continuous scale with decimal numbers. Computation is foreign to many scholarly disciplines and will remain so, since they formulate their problems in terms of language and thus explicitly allow vagueness in order to keep a larger range of possibilities in view. A fine language distinction could however be mapped in a rough way mathematically and thus open the door to ordering various alternatives. In my opinion, qualitative decision and game theory offers many possible applications even in the humanities.

- Normative decision theory tries to maximize the expected gain;
- Risk-gain analysis complements this maximization;
- Game theory takes the gain of each player into account;
- The "optimum" strategy may be poorer than a cooperative strategy.

4.3 Fuzzy Logic and Systems Control

Fuzzy logic has been developed since about 1980. It is used to formulate rules which allow machines to implement decisions for which only vague language instructions are available. Fuzzy logic, together with neural networks, are applied in particular in the engineering sciences. There, indefinite statements are associated with a particular area of applicability, and a membership function is defined to evaluate the relevance of the statements in a continuous, gradual manner. In Sect. 2.4 on linguistic indefiniteness, I defined membership functions and computed

[15]Cf. Lewis [11].

the corresponding indefiniteness. In dealing with indefiniteness in practice, we are confronted in contrast by the question: How can we establish rules for consistent action based on such vague statements?

The rule “If situation A applies, then carry out the action B” makes an unambiguous assignment of an action to a factual situation. But how can we apply this rule if it is uncertain as to whether A is in fact the case? Consider the following example: In a household, the space heating is to be regulated. To that end, the temperature t in a particular room is converted into a control factor y which regulates the heating unit.

If the room is warm (A1), turn down the heat (B1).

This first rule contains the vague statements A1 and B1. In order to arrive at a complete control process, we have to formulate a second rule:

If the room is cold (A2), turn the heat up (B2).

These two rules form a reasonable control process. But how do we deal with its indefiniteness? Fuzzy logic makes a suggestion about how to handle this problem. Each of the four statements (A1, A2, B1, B2) has a membership function as shown in Figs. 4.1 and 4.2.

The membership function for the statement “the room is warm” follows the perceived temperature as experienced by a “coddled Central European”, as described in Sect. 2.4; graphically, it takes the form of a triangle for temperatures t between 15 and 25 °C, with a maximum at 20 °C. For “cold”, we choose a corresponding function which ranges between 10 and 20 °C, with its maximum at 15 °C. These ranges guarantee a comfortable room climate and seem quite reasonable. Controlling the heating unit assumes some technical knowledge; its control factor y is presumed to vary between the values 0 and 15. In a similar manner to the way the opinions about the meaning of “warm” and “cold” overlap, the membership functions for the control factor will also overlap in a certain region where both the commands B1 and B2 apply.

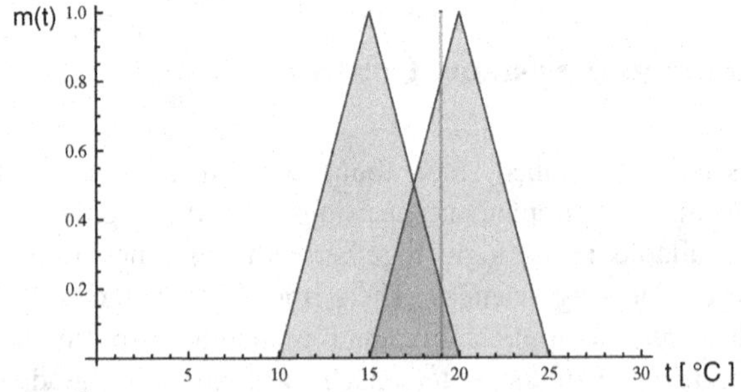

Fig. 4.1 The triangle on the *right* represents the membership function $m(t)$ as a function of the temperature (*horizontal axis*) 0 °C $< t <$ 30 °C for the statement A1 (“the room is warm”); the triangle on the *left* is the corresponding function $m(t)$ for the statement A2 (“the room is cold”). The *vertical line* indicates the current temperature (19 °C) of the room

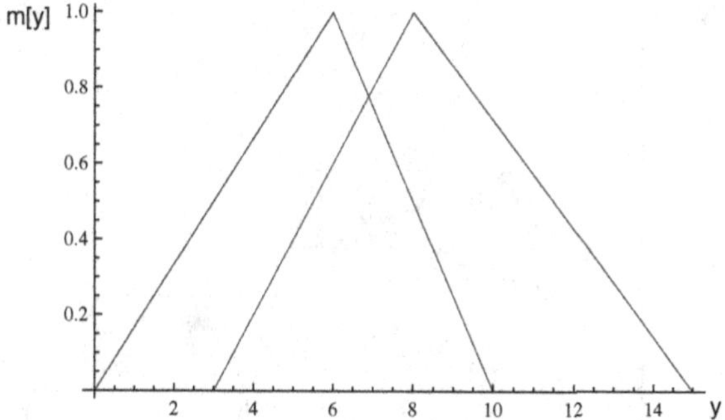

Fig. 4.2 The membership functions $m(y)$ as functions of the setting $0 < y < 15$ of the heating control factor y are shown for the statements B1 (*left*) and B2 (*right*). The opinion function for statement B1 ("turn down the heat") favors small values of y, while the opinion function for statement B2 ("turn up the heat") favors large values of y

I have chosen triangular, asymmetric membership functions due to possible delays in the response of the heating unit. Which control parameter must we choose for a certain input temperature that is neither warm nor cold, e.g. for the temperature t = 19 °C? From strict logic, we know that a rule of the form "if A, then B" is reasonable when both A and B are fulfilled. Fuzzy logic adopts this logical connection in the conjunction with the logical "and" symbol "∩". The two rules 1 and 2 are combined with the non-exclusive "or" symbol, the disjunction "∪":

$$(\mathrm{A1} \cap \mathrm{B1}) \cup (\mathrm{A2} \cap \mathrm{B2}).$$

The above set-theoretical statement is implemented in the membership functions according to the rules that we introduced in Sect. 2.4:

$$m[x;\ \mathrm{A} \cap \mathrm{B}] = \min\{m[x;\ \mathrm{A}],\ m[x;\ \mathrm{B}]\},$$
$$m[x;\ \mathrm{A} \cup \mathrm{B}] = \max\{m[x;\ \mathrm{A}],\ m[x;\ \mathrm{B}]\}.$$

Thus, in detail, the membership functions m_1 and m_2 for the two conjunctions at a temperature of t = 19 °C are given by:

$$m_1[y;\ (\mathrm{A1} \cap \mathrm{B1})] = \min\ \{m[t = 19\,°\mathrm{C};\ \mathrm{A1}] = 0.8,\ m[y;\ \mathrm{B1}]\},$$
$$m_2[y; (\mathrm{A2} \cap \mathrm{B2})] = \min\ \{m[t = 19\,°\mathrm{C};\ \mathrm{A2}] = 0.2,\ m[y;\ \mathrm{B2}]\}.$$

At the given temperature t, the membership function $m[y]$ can be evaluated for every value of the control factor for which this "fuzzy-logical" argument holds; when m = 1, we are required to "carry out unconditionally" the corresponding command, while for m = 0, we must "unconditionally avoid" the action. The resulting final function $m[y;\ (\mathrm{A1} \cap \mathrm{B1}) \cup (\mathrm{A2} \cap \mathrm{B2})] = \max\{m_1[y], m_2[y]\}$ has a chopped-off shape, as shown in Fig. 4.3. Since the present temperature of 19 °C lies more towards the higher end of the range, the membership function is higher for a

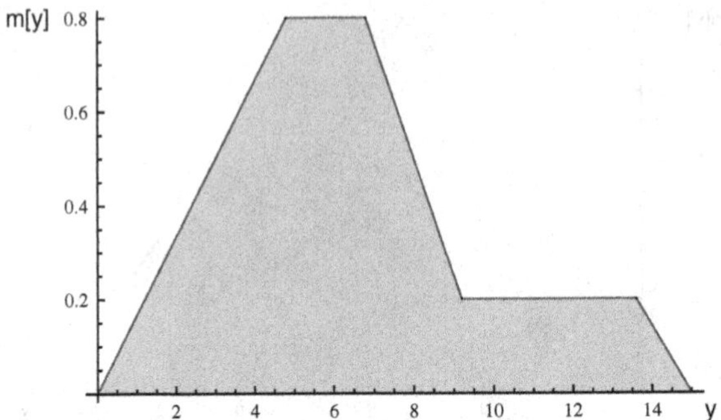

Fig. 4.3 The final membership function $m[y]$ (the vertical axis corresponds to the value of m, the horizontal axis to y), which associates a decision value m between 0 and 1 to every value of the control factor y ("pointer position") in the range $0 < y < 15$. The value $m = 1$ means "choose this pointer position unconditionally", while the value 0 means "avoid this value unconditionally". At the temperature $t = 19$ °C, the regulation of the system yields a pointer position which tends to favor lower values of the heating unit's output power

lower control factor, namely $m = 0.8$, while the larger value of the control factor yields a lower membership function of only $m = 0.2$. Now, there are several possibilities for converting this function into a pointer position ("defuzzification"). The most common is to compute the mean value from this weighting function:

$$\text{Pointer} = \int_0^{15} y m[y] dy \Big/ \int_0^{15} m[y] dy.$$

In the above example, we find for the pointer position $y = 6.4$. This example should be sufficient to explain the method. Every temperature is associated with a fixed pointer position corresponding to the given membership functions. What is the advantage of fuzzy regulation as compared to a lookup table with predetermined control parameters?

The membership function chosen here for the statement "the room is warm" corresponds to the subjective opinion of a particular person, an "expert", as described in Sect. 2.4. We could just as well have chosen a different opinion function. If then the second expert from Sect. 2.4 enters the house, he can choose his own membership function by pushing a button and the heating unit will respond to his personal preferences, using the corresponding pointer positions. Fuzzy regulation can thus readily translate a number of different, vague opinions of the users into commands for the machinery, in this case the heating unit. These variants can be indicated by symbols so that the users need not be aware of the details of the program. Such symbols ("landscape", "portrait", etc.) are employed for example by automatic cameras, allowing the camera to adjust itself optimally to the type of photo desired by making use of fuzzy logic.

A discussion of fuzzy logic must consider specific points as well as basic aspects. If one accepts the method, the important task is then the analysis of how to code the opinion functions. In order to program decisions for the regulation of complicated problems, one requires an expert whose judgment can be scrutinized. The user of the regulation program can criticize the opinion function of the expert. In practice, it is wise to test various expert systems by measuring their success rates. In the above example, success can be measured in terms of the comfort of the users and the energy consumption of the heating unit. A weak point of fuzzy logic is how the rules are determined; the number of rules increases with the number of variables that are to be regulated. The system then must optimize its rules by employing a neural network. The acceptance of fuzzy-logic methods is by no means universal. In Japan, "fuzzy technology" is employed in many places; the prime example is the subway system in Sendai. Its fuzzy-logic control has reduced the average transit time and the energy consumption for the subway system by 10 %. In economics, jurisprudence and in medicine, the application of rules when the prescribed specifications are ambiguous is an endlessly recurring problem. In medicine, computer programs offer both diagnosis and therapy recommendations based on unsharp judgments. Lists of simple, vague questions such as "Is the blood pressure too high?" or "Does the patient have an elevated temperature?", etc. can be integrated into a regulation system and can provide considerable improvement in the treatment of patients. The expert knowledge which is used to construct the regulation system is based on the statistics of many known cases.

Statistics depends on the probability theory of random events. Membership functions are not normalized and are therefore not probability distributions; they make no statements about whether one should assign vague elements to a certain class. Lofti Asker Zadeh has attempted to combine probability theory with fuzzy logic.[16] His approach is however regarded critically by Nozer D. Sinpurwalla and Jane M. Booker, who suggest instead a more complicated combination of vagueness and uncertainty.[17] They assume that "Nature" is exact and permits no limiting cases. This Nature is evaluated subjectively and judged objectively by an expert. The subject can then accept the judgment of the expert as a whole or only in part.[18] In my own opinion, this area is still a topic requiring active research.

In general, an algorithmic treatment of an indefinite case has both advantages and disadvantages for the user. Let us consider for example a physician: He obtains an independent second opinion by consulting an expert program, which helps him to check his own opinion. This is advantageous. However, the mathematical program reduces the urgency of further examinations and deliberations. It represents an ethical dilemma for the physician, since he is relying on a technology that he cannot completely understand. A new aspect in this example of a diagnostic program is the fact that here, software is intervening in a decisive way in the treatment of a patient.

---

[16]Cf. Zadeh [12].

[17]Cf. Singpurwalla and Booker [13].

[18]For more details, I refer the reader to the above reference.

We have in the meantime accepted the use of technical apparatus such as computers, whose inner workings we do not understand completely. But should we also allow life-determining decisions to depend on the workings of unknown software? A possible strategy would be to reject the expert judgment, as discussed in Ref. [13]—i.e. to bring the users and the programmers together for discussion. Then the users could continually improve the expert judgments in such programs on the basis of their own practical experience.

In the case of chess-playing computers, one can let different programs play against each other in order to determine empirically which one is the best. In contrast to the game of chess, where the rules are clearly formulated, the implementation of expert knowledge within a language of mathematical formulas is not so simple. The sciences of the mind often ignore the problems associated with control systems in the practical sciences. Instead, they limit themselves to a fundamental criticism of multivalued logic, which includes fuzzy logic.[19] They point out the basic nature of the decision between "true" and "false", and discuss the problems which arise out of a gradual choice between the two (see Sect. 2.4). In my opinion, no intermediate values between "true" and "false" are coded in the membership functions, but rather the indefiniteness of our knowledge. The definitions of the conjunction ("∩", i.e. 'and') and the disjunction ("∪", i.e. 'or') in fuzzy logic contain Aristotelian logic as a limiting case and therefore represent a reasonable extension. It is however significant for the relevance of the discussion of vagueness that in the mathematical theory (cf. Ref. [13]), an "exact Nature" without limiting cases has to be assumed in order to combine vagueness with uncertainty. The discussion of imprecise probabilities is an area which we have encountered before in our treatment of decision theory (Sect. 4.2).

In the *Marsilius Kolleg*, I experienced the objections of the humanities scholars to a quantitative analysis of vagueness and probability. In this context, the contribution of my colleague Andreas Kemmerling on vagueness independently of the state of information[20] is especially relevant. He emphasizes how important the problem is for fundamental rational thinking based on sound concepts. He argues that one should study empirically how different disciplines handle the problem. On the other hand, some of the participants held the view that in their fields, there is nothing to be quantified, nothing to be decided or regulated. But is it simply a prejudice of the natural sciences when they express the opinion that the humanities are missing out on a possible way of arriving at critical judgments by renouncing such quantitative methods?

- Fuzzy logic facilitates technical control;
- "and/or" are mapped onto minimum and maximum directives;

---

[19]Cf. Williamsen [14].

[20]Kemmerling [15].

- The pointer position for a heating unit is found from the mean value of the membership function;
- The coding of the membership functions is critical.

4.4 Finding the Right Metaphor

A common reaction to the indefinite is to present an analogy. The person or the group who cannot identify an indefinite object or concept A might make a statement like, "This A behaves in a similar manner to B", where B has well-known properties. The indefinite circumstances are transferred to a known situation, and this relocalization permits them to be classified within a definite sphere of knowledge or cognizance. In language, we transfer unknown situations into handy *metaphors*, which paraphrase the crux of the matter and bring it more vividly before our mind's eye. In the quote, "*Das Leben nennt der Derwisch eine Reise*" ("The dervish calls life a trip") from Heinrich von Kleist's "*Prinz von Homburg*",[21] traveling becomes a metaphor for life. Kleist fills the indefinite attitude towards life with the experience of traveling, which he takes to be more familiar to the reader than other kinds of experience.

What role do metaphors play in the (natural) sciences? Can they encompass essential aspects? Do metaphors open a new access to the search for truth? It is undeniable that metaphors are important for translating concepts from everyday language (natural language) into scientific terminology. The concept of "energy" means "drive" in everyday language, the ability to do something with vigor. In physics, energy refers to motion (kinetic energy) or to the latent ability to perform work (potential energy). The transfer of concepts has all the disadvantages that are well known from translations of one natural language into another. It expresses only a part of the original text correctly. Another example of this is the inadequate description of a quantum-mechanical object as a wave, which we found to be helpful for understanding the double-slit experiment (see Sect. 2.3). But for considering the energy and momentum balance in a collision, the particle picture is clearer than the wave picture. Furthermore, the physics of matter waves is qualitatively different from the physics of electromagnetic waves.

Are there metaphors which illustrate new perspectives? In the humanities, *hermeneutics* deals with questions of translation. The interpretation of a text produces a new text, which itself becomes an object of exegesis. Sometimes these interpretations can become circular: The image of an "hermeneutic circle" indicates that there is no linear path to the correct explanation of a text, but rather a gradual approach with incrementally increasing understanding. I believe that in the natural sciences as

[21] von Kleist [16].

well, the path to knowledge is often winding and that there is no essential difference from the process in the humanities. The natural scientist however may resolve that the iterations in this process must at some point be brought to a (preliminary) conclusion; then the problem has either been solved, or else it must remain unanswered. The classical areas of physics were considered to be completed, at least in a preliminary sense, by the late 19th century. Nevertheless, in mechanics, electrodynamics, and even in thermodynamics, there have been important newer developments such as chaos theory, electrodynamics in artificially curved spaces, or the dynamics of systems far from equilibrium, which indicate a lively continuing consolidation of these fields.

Does the use of "dark" metaphors in physics serve an hermeneutic purpose? The cosmologists have postulated the existence of "dark" matter and "dark" energy in order to bring newer observational results of astrophysics into agreement with Einstein's equations of gravitation for the expansion of the universe. They estimate from the masses of the spiral galaxies that the energy density of the visible universe makes up only 4.8 % of the critical mass density. The origin of the remaining 95.2 % of the energy of the universe is unknown. The attractive force of the visible stars in our galaxy is too small to hold its outermost stars in their orbits. Here again, one requires additional dark matter. It is called "dark" because it neither emits nor reflects nor absorbs light. The adjective *dark* is at the same time a metaphor, since the origin of the theoretical problems cannot be definitely attributed to new matter, because the corresponding elementary particles have yet to be discovered.

Like other matter, dark matter is subject to the gravitational force. For the very formation of the galaxies, such a mass of slowly-moving dark matter is also required in order to obtain the critical coherence for which the baryonic (nuclear, visible) matter alone is not sufficient. From the combined analysis of galactic clusters, supernovae and the microwave background radiation, limits for the percentage of cold, dark matter that contributes to the critical density of the universe can be set. From them, cold *dark matter* is found to make up 26.2 % of the critical density. Type Ia supernovae together with the microwave background radiation show that the remainder of the critical density must be attributed to *dark energy*, i.e. the dark energy contributes 69 % to the critical density of the universe. It is likewise invisible, since it does not interact with light, and it acts in the reverse sense to the attractive force of gravity, causing a repulsion which leads to an acceleration of the expansion of the cosmos.

The metaphors "dark matter" and "dark energy" are based upon the different relations between the energy density and the pressure in the equations of state of these two indefinite elements of the universe. Relativistic matter of light particles has an energy density which is three times larger than its pressure. Non-relativistic, heavy (*dark*) matter particles would have an energy density which results mainly from their mass density and not from their motion; i.e. their pressure is vanishingly small. The indefinite form of *dark energy*, in contrast, exerts a pressure which is equal to the negative of its energy density. These metaphors gave rise to the search for new particles or fields which would belong on the same side of the Einstein equations as the (known) baryonic matter and radiation components of the universe, in order to yield agreement with the astrophysical data. Parallel to the astrophysical research, particle physicists are therefore also involved in a search for new elementary particles which are heavy, interact only weakly, and thus appear dark (sometimes called "weakly-interacting massive particles" or WIMPs). For this search, experiments are carried out in tunnels deep in the earth, to avoid the disturbing effects of cosmic radiation from outer space.

As an alternative, one could of course also try to modify the other side of Einstein's gravitational equations, i.e. to revise general relativity theory. Independently of the metaphors of dark matter and dark energy, it has been questioned whether general relativity might in fact be in need of revision. Einstein himself added a cosmological constant to his original equations, and it could play the role of dark energy.

In addition to the four interactions that we presently know—the electromagnetic and weak interactions, the strong interaction and gravitation—there is perhaps a fifth force, the *quintessence*.[22] The term "quintessence" is a metaphor which originated with Aristotle, who added to the four elements of the Earth in antiquity a "*quinta essentia*", supposed to be found in the celestial sphere and to determine the eternal, changeless motion of the heavenly bodies. In contrast to the *quinta essentia*, the quintessence changes with time and would be responsible for the accelerated expansion of the universe as recently observed.

Natural scientists thus also use metaphors when they don't know exactly what a new discovery means. These metaphors refer to well-known concepts which can help to explain or understand the new discoveries. The resulting concepts are sometimes incorrect. Historically, the similarity of the electromagnetic wave equations to the equations for compressional waves or shear waves in elastic media was misleading. It seemed reasonable to postulate an *ether*, which would serve as medium for the propagation of electromagnetic waves. Similarly to an elastic material, the ether would provide a medium which could carry waves by being compressed or distorted. Only after the experiments of Albert A. Michelson and Edward Morley, and the special relativity theory of Albert Einstein, was this transferral revealed to be inappropriate.

A philosopher might interpret the "neural network" as a metaphor for the complex grey matter of our brains, or for individual neurons within it. A biophysicist would object and say that a that a neural network is a *model* for neuronal processes. When is a neural network a real model? In certain equations for neural networks, measurable excitation potentials occur. In the Hodgkin-Huxley model of the action potentials in the nervous systems of squids, for example, the ionic mechanism is described in detail; it plays out during excitation and inhibition in the peripheral and central regions of the nerve-cell membranes. In this model, the voltage-dependent resistances and capacitances are simulated and can be compared with data from experiments on the ion channels. We would therefore consider it to be a physical model.

[22] Cf. Wetterich [17].

A theoretician has to analyze his data formally. An important aspect in judging different theories is the beauty or elegance of a particular theory. Are the results obtained from many assumptions and approximations, from repeated approaches applied under a variety of circumstances, or is the theory convincing because of a simple and aesthetically-pleasing derivation? Reductionism, with its striving for simplicity, is so attractive because the simple is beautiful. Transferring unrelated numerical empirical data into an "elegant" equation confirms the researcher in his hope of having discovered an important relation. But even an elegant theory can be wrong.

Hans Blumenberg has dealt in detail with the history of metaphors and their different roles.[23] In particular, he investigated how metaphors arise in the neighborhood of new knowledge. The discovery of the heliocentric worldview in astronomy in his view triggered a process of "self-diminishment of humanity". But it would also be possible for humans to celebrate this perception as a triumph of the human mind. Similarly, considerations of the indeterminacies in the natural sciences can qualify or devalue the knowledge that has been acquired. The limits of our present knowledge could cast doubts on the value of continued research. This would be a position of the sort that is not unusual in so-called post-modern discussions. "Science can never be a mirror of nature; there are no neutral observers, all experiments are theory-laden, there are no simple facts".[24] Babette E. Babich advances the thesis that the so-called post-modern turn, the emphasis on the indefinite and informal in physics, is not sufficient to qualify the findings of the natural sciences. A reasonable philosophy of science would have to carry out a serious debate including the criticisms of Heidegger, Husserl and others. Edmund Husserl claims that the empirical sciences cannot account for their own normativity. A different philosophical stance is called for, which not only describes the world as it is, but shows how it ought to be understood. "The exclusive worldview of modern man in the second half of the 19th century, based on the positive sciences and blinded by the prosperity arising from them, means that men turn away from those questions which are decisive for humanity".[25] One of Babich's hypotheses is that the successful treatment of indeterminacy in the sciences confirms their claim that they have come close to the reality of nature. But only when science fails are social causes advanced in order to avoid recognizing the universal relativism of scientific truth. A topical example of this could be the international fusion project "ITER", which is well behind its planned schedule. The concept itself is however never questioned, but instead the coordination and planning for the construction of the project. Other advocates of social epistemology in the philosophy of science assert that theories survive for sociological reasons and that they therefore contain no innate reality. This approach fails to recognize that well-established theories are accepted in the most diverse societies. Even in the

[23]Cf. Blumenberg [18].

[24]Cf. Babich et al. [19].

[25]Husserl [20].

ideologically petrified Soviet Union of the Stalin era, quantum mechanics was still taught in the universities.

- Transferral of the indefinite into a familiar context;
- Transferral in language makes use of metaphors;
- Even an elegant theory can be wrong;
- Dealing with indeterminacy does not limit the value of science.

4.5 Following the Signs

In order to determine the indefinite, one requires a structured group of classes to which the indefinite concept or object can be assigned. One could thus assume a relation which identifies the indefinite object with a certain subset of a system. This assumption however overlooks the person who is carrying out the classification, or rather, it overlooks the *reason* for the assignment. It is also too limited, since the indefinite object in fact plays two different roles; namely as an individual case, and as an example for other indefinite phenomena of a similar kind. Does the indefiniteness arise from the uniqueness of the phenomenon or from an outmoded classification of known phenomena? Is a change of perspective or an extension of the list of known subsets necessary?

In the practical approach which I prefer, the significance of an indefinite event in relation to a particular model or theory is of major importance. There are indefinite results whose significance is not in urgent need of updating, since it is not very clear to anyone what might be gained from more precise investigations. The responsibility of the experimenter should be emphasized, as we have already pointed out in Sect. 4.1. Indefiniteness can simply be uninteresting, as has been demonstrated by the established institutes for the study of ESP, i.e. extrasensory perception. Even if a test person can guess the identity of a hidden playing card with a higher probability than expected, this "telepathy" need not necessarily be investigated scientifically. The situation is different in the case of results which would imply an important paradigmatic change in the development of theory. Here, every scientist will be eager to verify and improve the result as quickly as possible and as precisely as possible in order to find the decisive answer. We as scientists should however be aware that we often apply subjective criteria in choosing relevant topics of research.

In physics, the concept of atomic weight is not at all indefinite, but the measurement of atomic weights can yield non-categorizable limiting cases, as shown in Sect. 4.1. One has to identify the new category, in this case the category "neutron number", in order to explain the indefinite differences between expected and measured values. In the case of Hamiltonian random matrices (see Sect. 2.1), and in the treatment of stochastic differential equations, the intention is to lend structure to our lack of knowledge by constructing symmetry classes of random matrices which will then correspond to similar classes of phenomena. Likewise, in modeling biological systems, one can discern the intention of finding generic classes of models which will offer explanatory patterns without requiring knowledge of numerical parameters.

In the natural sciences, new data can be integrated into a known theory when its determining elements are sufficient to analyze them and perhaps later to explain them. Pierre Duhem compared a physical experiment with everyday experience and found that experiments are less certain but much more precise than the non-scientific report of an occurrence.[26] A witness to a traffic accident may remember the directions in which the autos involved were moving and their speeds, the time of the accident and where he or she was standing, how loud the crash was, etc. He may be able to recount a number of anecdotes which he insists accurately reflect the course of events. If an experimenter is asked about the results that his apparatus has produced on a certain day, he can consult his laboratory logbook and will find for the most part columns of numbers, in addition to general remarks on the condition of the apparatus. In modern times, he will have stored the results from various different experiments digitally, so that he can reconstruct the recorded events precisely. If, however, he were required to reconstruct them in everyday language, like the witness to an accident, his report would be much more confused and uncertain. Simply as a result of limiting the data to those whose interrelationships can be treated by mathematical theory, his description can attain the high degree of precision that we expect from physical measurements.

History takes on an intermediate position between the positive natural sciences and the interpretive humanities. There are historical facts which the historian can determine by consulting original sources. This can lead to indefiniteness, for example because the language has to be translated, e.g. from Latin or a Medieval dialect into modern, 21 century language. Or there are contradictory sources; an example is the "clash" between Lothair of West Francia and the Holy Roman Emperor Otto II in Aachen in the year 978.[27]

The monk Richer reports: "...the bronze eagle[28] [...] was turned around [*by Lothair's troops*] and pointed to the East, for the Germanic troops had turned it to the West, as a subtle way of indicating that their army could defeat the Gauls." Bishop Thietmar relates the same story as follows: "King Lothair, who with his strong army [*had ventured*] to occupy the imperial palace and the king's throne in Aachen, and to turn the eagle towards himself [...]. It was the custom that anyone who took possession of this place would turn it towards his realm."

Since Lothair's troups had come from France, Thietmar's source would imply that the eagle was turned towards the West, opposite to the direction affirmed by

[26]Cf. Duhem [21].

[27]Cf. Schneidmüller [22].

[28]To understand the significance of the eagle, one must be aware that Charlemagne had made an eagle the symbol of his empire (later the Holy Roman Empire).

Richer. The western orientation asserted by Thietmar is used to support Lothair's claim to Aachen, while the eastern orientation attested to by Richer represents the overcoming of an historical predominance of the Germanic nation. The work of the historian is to understand these sources in relation to their contextual interpretations. Perspectives can change. In the 19th century, the German-French conflict was interpreted in the context of the national aspirations of the European countries; in the present-day European context, one sees a process of ethnological differentiation. "The postmodern belief in the nearly unlimited power of historians to shape the interpretations of the past is evidently a consequence of the view that an historic source [...] can be seen as nothing more than the carrier of a fixed, unchangeable meaning".[29]

As in the Platonic discussion (Sect. 2.6), I would suggest here, in addition to the indefinite and the definite, to bring a third important element into play: it is the "*cause*". What goals were the two witnesses to the historic clash between Lothair and Otto pursuing in their two different portrayals? For them, the eagle was the symbol of a political intention. This sign was set in order to document political power. Semiotics, the theory of signs, analyzes such relationships in languages, texts and societies. It organizes the indefinite and the determining, and the cause which brings them together. The structuralistic triad consisting of the indefinite object, the determining interpretants and the constitutive sign is very useful for elucidating the process of determination. We will explain this triad using two examples from physics and biology, and in particular we will ask about the *cause* for the union of the indefinite and the definite.

Up to now, it has seemed that we must count on the indefinite. "*Count*" is meant here in two different senses: On the one hand, count refers to mathematical methods for quantifying the indefinite, and for integrating it by degrees into a body of rules. On the other hand, I wished to emphasize that one must take indefiniteness *into account*. For determinations, the standards are higher, since now the pair of opposites, indeterminate and determined, must be brought into a justifiable relation to each other. The metaphor (Sect. 4.4) foreshadows such a relationship, while the sign justifies it.

The theory of signs, semiotics, arose historically from linguistics and ethnology. The significance of symbols and signs in the natural sciences can be traced back to Herman v. Helmholtz and Heinrich Hertz.[30] Starting from physiological phenomena, the sign was introduced as a neural complement to physical experience. Sounds are perceived not simply in terms of their intensities as measured by the physicist; instead, the brain evolves its own perceptions of harmony or dissonance. Symbols are more than copies or images of the physical in the mathematical world. Ernst Cassirer saw the symbol as the center and reference point for physics and epistemology.[31] I prefer the concept "*sign*", since in general, symbols

[29]Evans [23].

[30]Dosch [24].

[31]Cassirer [25].

are more difficult to understand than signs. In contrast to signs, symbols are intrinsically bound up with a person or a group of persons, who share the same nationality, culture or environment; thus there is no single dictionary, but rather many different dictionaries for symbols. Signs are simpler; they transmit information from our daily lives. Traffic signs for example have international validity and unique meanings. For a pragmatic analysis of the natural sciences, signs would therefore seem to be more expedient. In particular, signs which stand for mathematical operators in equations that describe natural phenomena have proved their worth.

The triad of the indefinite, the definite and the cause of their union is more dynamic than the static triad consisting of object, interpretant and sign. It reflects the process of *determination*. Traditionally, the "*sign*" is defined by Charles Sanders Peirce in his "Syllabus of Certain Topics of Logic" as follows: "A sign is everything which is related to a second thing, which is called its object, in such a way that the sign can determine a third thing, which is called its interpretant, to be related in the same triangular relation to the object, as the sign is related to the object".[32] I would interpret this complicated sentence as follows: The indefinite object must be determined by the data (interpretant) and justified as a sign in a larger context. Peirce takes the route which starts from the sign and moves on to the object.

The terminology can be most simply explained using an example from classical physics. Let us take as object an apple which falls from a tree. The observer wishes to determine the path of the apple during its fall; he therefore studies the distances traveled by the apple during fixed time intervals using a fast camera. If the observer is interested not only in falling apples, but also in falling pears, he becomes an experimenter. He records similar images of falling pears and compares the coordinates of the distances covered per unit time from the various images. Since the two paths have a similar, parabolic form, he recognizes that they don't depend on the type of fruit observed. He thus defines falling objects independently of their shape and form, and treats them as pointlike, lacking color, taste or form. He constitutes the indefinite object as an object with only one property, its weight, which is related to the force of gravity. He calls this object a point mass, and postulates that its motion follows from a natural law. This law is expressed by the equation of motion in which the coordinate $x(t)$ denotes the position of the massive object. The sign $x(t)$ is integrated into an explanatory mathematical relation, namely the equation of motion $m\ddot{x} = mg$, in which the left-hand side symbolizes the (inertial) force of acceleration and the right-hand side the force of gravity. The cause of the motion is the acceleration of gravity g, which makes the point mass fall (cf. Fig. 4.4).

The sign becomes part of a sign language which in the physical sciences is mathematics. Conversely, the sign determines the interpretants, that is the data which are generated by the indefinite object. The interpretants cannot add anything to the sign which was not already present, for example they can add no information about the temperature of the object.

[32]Cf. Peirce [26].

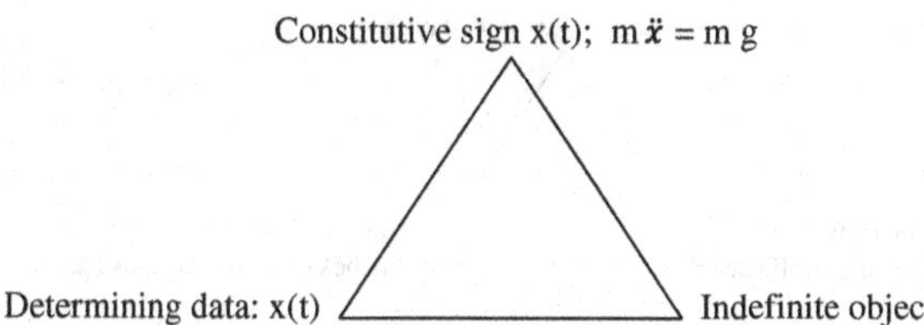

Fig. 4.4 The triad consisting of the indefinite object, the determining data and the constitutive sign is illustrated by the example of the law of falling bodies. It combines the distance fallen and the time (phenomena) with the apple (real object) and the Newtonian equation, mass × acceleration = force of gravity (ideal relation)

Often, the "cause" follows the determination. But they need not necessarily appear in that order. In present-day physics, a lively discussion about string theory is underway. String theory is a multifaceted theory that has been developed over roughly forty years, which has however not been found to be identifiable with any physically-measurable properties of objects. Here, the sign is thus isolated from the still-to-be-determined data and objects. Strings represent a generalization of quantized, pointlike particles. In the simplest case, they are two-dimensional filaments which draw out a carpet-like surface in spacetime, rather than the worldline along which pointlike particles move. Only in the past ten years has the correspondence between a particular string theory with anti-de Sitter geometry and a Yang-Mills theory been postulated as a result of the conjecture of Juan Maldacena.[33] It should be very similar to the quantum dynamics of the quarks, i.e. the hadronic elementary particles. This theory evolved not inductively, but rather deductively from purely mathematical principles. To what extent this development will prove to be fruitful remains to be seen, but it represents a good example of a different temporal ordering.

Biological examples are important in order to demonstrate the process of determination in its greatest generality. They are not directly connected with mathematical symbols. The Darwin finches on the Galápagos Islands cannot be attributed to a certain family (*incertae sedis*).[34] Darwin explained the great variety of finches on those islands in his "Origin of the Species" as follows: "…everything is incomprehensible under the assumption of an independent creation of the species, but not under the assumption that they migrated from the nearest or most suitable source, with subsequent adaptation of the colonists to their new home".[35] Let us consider the particular species *conirostris* among the 12 different species of the genus of the cactus finches (*geospiza*). Finches of this species exhibit two types of song and two forms of beaks, which are longer or shorter, with which they extract the cactus fruits in different ways. This double form maximizes their options for

[33]Cf. Maldacena [27, 28].

[34]Cf. Remsen et al. [29].

[35]Darwin [30].

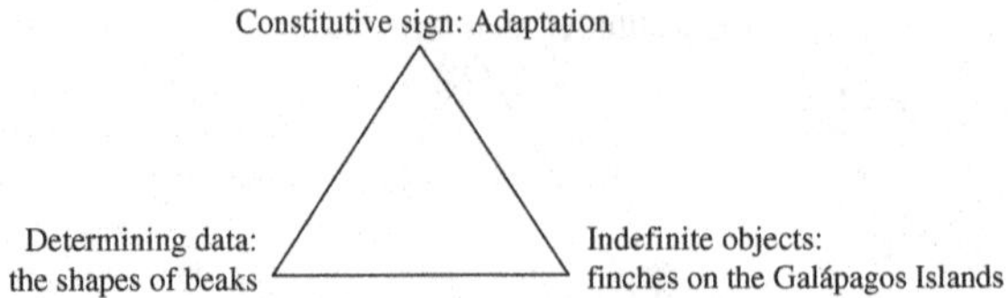

Fig. 4.5 The triad consisting of the indefinite, the determining, and the "cause" of their union, illustrated using the example of the Darwin finches. The shapes of the beaks (phenomena) of the finches on the Galápagos Islands (real objects) are related through adaptation to their environments (ideal relation)

feeding. Inspired by Darwin's ideas, Rosemary und Peter Grant studied these birds in detail for 30 years.[36]

Turning back to our question of the Darwinian triad, I will consider the finches that belong to no particular family to be indefinite objects. The determining data are their different beaks and feeding habits. Why are there so many different species? Is there a cause for this?[37] Optimal adaptation to their environments would seem to be a possible explanation.

The example has been investigated at the molecular-biological level only recently.[38] A gene has been identified that stimulates production of the protein *calmodulin*, which is present at higher concentrations in finches with longer beaks.

Adaptation has been identified as the explicatory cause in the particular example of Darwin's finches. In Fig. 4.5, the triad—the sign—is represented with its three components. Adaptation is a sign for evolutionary theory.

Can we mix the indefinite with the definite in such a way that new causes become apparent? In the process of explanation described above, the determination plays an important role. Biologically, it is a specific property of the finches which becomes the determining element. Just so, a chosen property of the falling objects, namely their position coordinate, characterizes Newtonian dynamics. From the falling apple to the meaning of the mass in gravitation is still a long path. Similarly, the meaning of adaptation for evolution has been clarified only retrospectively. "The determination supports the explanation and the explanation guarantees the intervention and the utilization",[39] according to Robert Klee, who prefers medical-biological rather than physical examples in his philosophy of science. Concepts must be all the more clearly formulated, the less applicable a mathematical description becomes. The sign must completely describe an aspect of the indefinite object in the empirical sciences. However, it is just as important that the scientist describes the situation in a linguistically clear manner, that he develops it logically

[36]Cf. Grant and Grant [31].

[37]Here, I use *cause* and *justification* synonymously, although not every justification also represents a true cause.

[38]Cf. Patel [32].

[39]Klee [33].

and penetrates to its core. If these criteria are fulfilled, scientific investigation can deliver a true, i.e. an adequate description.

- Indefinite objects, determining data, and the cause of their union;
- Determination as the setting of signs;
- Examples from Newtonian mechanics and Darwin's origin of species;
- Determination supports the explanation.

4.6 Actors and Scenarios: Theater Instead of Theory

In scientific research, a large part of the thinking and argumentation takes place in the form of *theory*. Etymologically, the word "*theoria*" means *viewing*, derived from the ancient Greek words "*thea*", view, and "*horan*", to see. The word *theater* is derived from the same stem, "*thea*". In the theater, our view is directed towards the stage on which actors are presenting a play. One could thus follow David Bohm in saying "theory is primarily a form of *insight*"[40] and not a form of knowledge. The language of the theater, with the concepts "actor" and "scenario", is employed especially in the "new" sociology and in ecology, which is closely connected with politics and society. In discussing inanimate Nature, we have used the term *agents* (see Sect. 3.5); they pursue their interactions in a passive manner. In contrast, human beings think and act independently within society. Therefore, in sociology the concept of the *actor* has been developed. The sociologist Bruno Latour makes use of this approach. He deals with indefiniteness in a different way from the natural scientists, but he uses a terminology which is in part derived from concepts used in the physics of systems, of networks and order–disorder phenomena. I wish now to discuss this new approach and compare it with the structural method of determination described in Sect. 4.5.

As one of the protagonists of this theoretical development, Bruno Latour in his "Actor-network theory" places the person who acts and develops himself as an equal alongside the network.[41] A network is something purely technical and mechanical, and the term gives only an inadequate description of the object that is meant here. Actors are the persons who deal with the network. In Sects. 3.5 and 3.6, we described the theory of complex networks, in which networks make up the determinate, in contrast to their indefinite surroundings. Bruno Latour assumes that the objects (agents) of the network can become independent, so that they themselves become actors. What does this mean? The earthquake in Japan in the year 2011, for example, set in motion a catastrophe which destroyed several nuclear

[40]Cf. Bohm [34].

[41]Cf. Latour [35].

reactors at the Fukushima power plant. But an analysis of the internal technical structures of those reactors gave no evidence that such a result could have been expected. The natural catastrophe was in this sense a constitutive force, which set in motion reflection and strategies for action. In Germany, for example, this accident provoked a revision of official policies regarding nuclear energy. Bruno Latour therefore maintains that non-human entities can "act" independently. "An actor [...] is something that acts, implies no special motivation of human individual actors, nor of humans in general [...]. The self and the social actor of traditional social theory is not on its agenda (i.e. of the Actor-network theory)".[42] In my opinion, one can attribute to natural causes the ability to prompt to action. But the concept of "acting" should best be reserved for human beings. However, I agree with Bruno Latour that society and Nature do not exist separately.

Humans are a part of Nature, and Nature changes our social organization. One clearcut example: Global warming is changing the ecological system of the oceans and their populations of fish. This is having a drastic effect on fisheries worldwide. "Just as much as Nature gives to humans, humans should give back to Nature, which has become a subject of justice",[43] asserts the philosopher Michel Serres. Animals and plants should have votes in the democratic process, according to his opinion, and should be represented by shamans who can communicate with them.

The separation of system and environment—or of the determinate and the indefinite—is nullified here, which implies far-reaching consequences for sociological research. Organizational problems in choosing the proper shamans can also be expected.

Bruno Latour as a philosopher of science is known for his study in collaboration with Steve Woolgar on "Laboratory Life", in which the authors describe a project at the Salk Institute in California, where hormones that act on the nervous system were under investigation.[44] They see the laboratory as a kind of "*terra incognita*", and the scientists who work there as an alien tribe of natives. For the two sociologists, the biomedical scientists are outsiders who confront them with foreign concepts. Bruno Latour and Steve Woolgar are interested mainly in the compilation and writing of scientific publications ("texts") and not in the models and experiments which underly them. In this sense, their method is closely related to semiotics, which also starts from an isolated text, without investigating its context and history. I agree with them in their extension of the semiotic theory[45] to the objects of the natural sciences: "In the practice of Actor-network theory, semiotics was extended to define a completely empty frame, that enabled to follow any assemblage of heterogeneous entities—including now the natural entities of science and the material entities of technology." The sociologist evolves a sort of exposé within a dense description, whose performers are the *actors*. In the interlude "What should be done with the Actor-network

---

[42]Latour [36].

[43]Michel Serres [37].

[44]Cf. Latour and Woolgar [38].

[45]Siehe Referenz [36].

theory?",[46] a student asks his professor: "Just describing. Pardon me, but isn't that awfully naive?" The professor replies, "Why do you think that describing is simple? You are no doubt confusing it with compiling a long list of clichés. [...] Describing, being aware of the concrete circumstances, finding the uniquely adequate report for a given situation, I have always found that to be exceedingly exacting."

But yet, the method of investigation inspired by Bruno Latour links up a chain of interpretants, without explanations or causes; indeed, it keeps those causes hidden, because the actions of the actors within their interrelations presumably explain everything. Bruno Latour's method can even completely miss out on the kernel of the matter.

In his semiotic analysis[47] of Albert Einstein's introductory text on special relativity theory,[48] Latour reports on the important actors, their actions, and the metaphors which occur in this text. The elements of special relativity are generalized in a discussion and applied to the social sciences (p. 26 ff). The article distinguishes between relativism and relativity, but it fails to comprehend the first postulate of special relativity theory, i.e. that the laws of physics should have the same form in all inertial systems.[49] Since Bruno Latour refers only to Einstein's text, its relation to reality escapes him. When certain relativistic invariants that are formed from the energies and the momenta of two colliding protons coincide, the scattering of a fast proton off a proton at rest gives the same result as an experiment with two colliding protons that are moving in opposite directions. This invariance has nothing to do with Einstein's "conservative" thinking. Likewise, the second postulate, the invariance of the velocity of light, independently of the motion of its source or of the observer, is suppressed in Latour's analysis of Einstein's text. Bruno Latour's considerations are fixed upon a "center of calculation" as the hub of a network which is connected through measurements ("metrologically") with the satellites. This seems to me to be something like a scientific fiction: Einstein's theory is regarded as a metaphor which can be transferred to sociology.

I am convinced that complex systems can be better understood with the aid of networks. For this reason, it is all the more important to me to emphasize the differences with regard to Latour's semiotic approach; compare the above with Sect. 4.5. A system of knowledge consists not only of texts.[50] The interpretants are often experiments and observations, and the objects are often real entities. Physics penetrates a problem successfully when it can reduce several experiments to a single property which mutually interprets them. In mechanics, in the case of a freely

---

[46]Cf. Latour [39].

[47]Cf. Latour [40].

[48]Cf. Einstein [41].

[49]Inertial systems are coordinate systems which move relative to one another at constant velocities. The typical example in Einstein's writings is a train which is traveling on a straight track at constant speed, compared to a 'motionless' observer on the station platform.

[50]The two examples in Sect. 4.5 are taken from the natural sciences; naturally, the humanities often investigate objects which are available exclusively in the form of written texts. The tryptich of interpretant, object and sign seems to me nevertheless to be useful for determination in these fields.

falling object, this reduction consists of the postulate that the objects have an inertial and a gravitational mass, as described in Sect. 4.5. In the special theory of relativity, the invariance of physical relations under Lorentz transformations and the constancy of the velocity of light are also such postulates.

The interpretants, e.g. the data on the distance fallen as a function of the time, say only very little about the falling apple; they say nothing about its taste or its color. If a more complete picture of the situation is required, one needs to add more information. Reduction does not suffice when a scientist is trying to establish connections between different fields at their boundaries; he then has to combine the attributes of the one field with those of the other field. Physics concentrates on structures in order to investigate the organization and composition of matter; biology is interested in functionalities in order to fathom the working principles of microscopic units within whole living systems. Ecology has to take into account ethical and judicial aspects which can create environmental problems. Therefore, at the boundaries of these sciences, exciting questions arise that deal with the structures, the functions and the values of living Nature.

Scenarios are another scientific form in which many newer studies present themselves at the border regions between the natural sciences and social theory. When the natural environment becomes unpredictable owing to human intervention, it becomes necessary to plan the various possible reactions of humanity in advance. In its fundamental study, "The Limits of Growth",[51] the Club of Rome in 1972 described the growing industrial production, the increasing global human population, and the resulting possible collapse of the world's economy. Due to the finite amount of resources, this is a thoroughly realistic development. The authors of the Club of Rome's report dealt with this topic in terms of different scenarios: What would happen if the amount of available resources were doubled? How does the model change if an unlimited amount of energy becomes available, e.g. through nuclear fusion? How can a collapse be delayed by controlling environmental pollution? Is there a possibility of sustainable development? The suggestions of the Club of Rome are indeed still relevant today: control of population growth, reduction of the consumption of resources, shifting of economies from production to services, education and health. Similar scenarios for projections of the increase in global CO_2 concentration have been proposed. The scenarios of the IPCC Report 2000[52] contain four "storylines" with four different themes; these are rapid general industrial development (A1), heterogeneous independent development (A2), global development towards service and information technologies (B1), and sustainable development (B2). Each of these families contains an additional three variants (A1FI, A1T, A1B) within a harmonizing (HS) and an open assumption (OS) about the growth rates of population, economy and energy consumption. All together, 40 scenarios were considered. The differences in the predictions for 2010 were drastic. The study recommends that the various scenarios be mutually deliberated upon and

[51]Cf. Meadows et al. [42].

[52]IPCC [43].

that political decisions be made on the basis of their comparison. Is this feasible with such a multiplicity of scenarios? Or have the scientists avoided all further political responsibility by providing such a detailed consideration of possible future developments?

Scenarios should be seen as variations on theories and not as theoretical speculation. Naturally, scenarios convey the message that here, science is making somewhat less stringent prognoses than is usual in scientific treatises. The number of important parameters is indeterminate, and new ones may enter from scenario to scenario. The relevant laws are known only in part, and often, extrapolations of functional relationships must take their place. Nonlinearities strongly influence the time variation of various important quantities. All of these limitations lead to the result that the error limits of such studies are hard to estimate. Nevertheless, scenarios offer a possible alternative to a theory whose validity cannot be supported by a thorough analysis of systematic and empirical errors.

- Society as a network of independent agents;
- A contract with Nature, in order to protect her;
- Scenarios for estimating the effects of human interventions;
- A scenario is better than a theory without error analysis.

References

1. Rheinberger, H.-J.: Man weiss nicht genau, was man nicht weiss. Über die Kunst das Unbekannte zu erforschen. Neue Züricher Zeitung 5./6. Mai 2007, S.30 (2007). http://www.nzz.ch/2007/05/05/li/articleELG88
2. Cassirer, E.: Philosophie der symbolischen Formen, Band 3: Phänomenologie der Erkenntnis, Dritter Teil: Die Bedeutungsfunktion und der Aufbau der wissenschaftlichen Erkenntnis, Darmstadt, p. 357 (Philosophy of Symbolic Forms (1923–29), English translation 1953–1957) (1994). http://www.arts.rpi.edu/ruiz/AdvancedIntegratedArts/ReadingsAIA/Cassirer%20Toward%20a%20Theory%20of%20the%20Concept.pdf
3. Osheroff, D.D.: Superfluidity in ^{3}He: discovery and understanding. Rev. Mod. Phys. **69**, 667–682 (1997)
4. Barber, R.C.: Isotopes. In: Lerner, R.G., Trigg, G.L. (eds.) Encyclopedia of Physics. Addison-Wesley, Reading (1981)
5. Gamm, G.: Flucht aus der Kategorie – Die Positivierung des Unbestimmten, Frankfurt am Main (1997)
6. Authier, M.: Elemente einer Geschichte der Wissenschaften, p. 17. In: Michel Serres (ed.) Frankfurt am Main (1997)
7. Taleb, N.N.: The Black Swan—The Impact of the Highly Improbable, p. 119. Random House, London (2007)
8. Dörsam, P.: Grundlagen der Entscheidungstheorie, Heidenau, p. 39 (1996)
9. Smith, J.M.: Evolution and the Theory of Games. Cambridge (1982)
10. Nash, J.F.: Non-cooperative Games. Dissertation, Princeton University, Princeton (1950)

11. Lewis, D.K.: Convention—A Philosophical Study. Harvard University Press, Cambridge (1969)
12. Zadeh, L.A.: Probability measures of Fuzzy events. J. Math. Anal. Appl. **23**, 421–427 (1968)
13. Singpurwalla, N.D., Booker, J.M.: Membership functions and probability measures of fuzzy sets. J. Am. Stat. Assoc. **99**(467), 867–877 (2004)
14. Williamsen, T: Précis of Vagueness. Philos. Phenomenol. Res. **57**(4) (1997)
15. Kemmerling, A.: Informationsimmune Unbestimmtheit. Bemerkungen und Abschweifungen zu einer klaffenden Wunde der theoretischen Philosophie, 40 pp., Online-Journal Forum Marsiliuskolleg. doi:10.11588/fmk.2012.0.9407
16. von Kleist, H.: Der Prinz von Homburg, 4th act, third scene. Hamburg (1960)
17. Wetterich, C.: Quintessenz – die fünfte Kraft. Welche dunkle Energie dominiert das Universum? Physik Journal **12**, S. 43 ff (2004)
18. Blumenberg, H.: Paradigmen zu einer Metaphorologie. Frankfurt am Main, S. 142 ff (1998)
19. Babich, B.E., Bergoffen, D.B., Glynn, S.V.: Continental and Postmodern Perspectives in the Philosophy of Science, Aldershot, Brookfield (1995). http://www.fordham.edu/gsas/phil/babich/cppref.htm
20. Husserl, E.: Die Krisis der europäischen Wissenschaften und die transzendentale Phänomenologie, p. 7. Hamburg, 2012 (The Crisis of European Sciences and Transcendental Phenomenology: An Introduction to Phenomenological Philosophy). Northwestern University Press, Evanston (1970)
21. Duhem, P.M.M.: La théorie physique: son objet, sa structure, p. 246, deuxieme edition 1914. Paris (1981)
22. Schneidmüller, B.: Die Begegnung der Könige und die erste Nationalisierung Europas (9.–11. Jahrhundert). In: Le relazioni internazionali nell'alto medioevo, Settimane di studio del Centro Italiano di studi sulla alto medievo 58, Spoleto 2011
23. Evans, R.J.: Fakten und Fiktionen, p. 104. Über die Grundlagen historischer Erkenntnis, Frankfurt am Main (1999)
24. Dosch, H.G.: The concept of sign and symbol in the work of Herman Helmholtz and Heinrich Hertz. In: Janz, N. (ed.) Cassirer 1945–1995, pp. 47–61. Sciences et Culture, Lausanne (1997)
25. Cassirer, E.: Philosophie der symbolischen Formen, Band 3: Phänomenologie der Erkenntnis, p. 25. Darmstadt (1954)
26. Peirce, C.S.: Syllabus of Certain Topics of Logic. Boston (1903)
27. Maldacena, J.M.: The large N limit of superconformal field theories and supergravity. Int. J. Theor. Phys. **38**, 1113–1133 (1999)
28. Maldacena, J.M.: The large N limit of superconformal field theories and supergravity. Adv. Theor. Math. Phys. 2:231–252 (1998). e-Print: arXiv:hep-th/9711200 (1997)
29. Remsen, J.V., Jr., Cadena, C.D., Jaramillo, A., Nores, M., Pacheco, J.F., Robbins, M.B., Schulenberg, T.S., Stiles, F.G., Stotz, D.F., Zimmer, K.J.: A classification of bird species of South America. American Ornithologists' Union (2007)
30. Darwin, C.: The Origin of the Species, p. 571. Stuttgart (1963)
31. Grant, B.R., Grant, P.R.: Evolutionary Dynamics of a Natural Population: The Large Cactus Finch of the Galápagos, p. 241. Chicago (1989)
32. Patel, N.H.: Evolutionary biology: how to build a longer beak. Nature **442**, 515–516 (2006)
33. Klee, R.: Introduction to the Philosophy of Science: Cutting Nature at Its Seams, p. 25. New York and Oxford (1997)
34. Bohm, D.: Wholeness and the Implicate Order, pp. 4–5. Routledge, London (1980)
35. Latour, B.: Reassembling the Social: An Introduction to Actor-network Theory. Oxford (2005)
36. Latour, B.: On actor-network theory. A few clarifications (1996). http://www.nettime.org/Lists-Archives/nettime-l-9801/msg00019.html
37. Serres, M.: Le contrat naturel, p. 67. Paris (1990)
38. Latour, B., Woolgar, S.: Laboratory Life: The Construction of Scientific Facts. Princeton (1979)
39. Latour, B.: Eine neue Soziologie für eine neue Gesellschaft, p. 244. Frankfurt, Einführung in die Akteur-Netzwerk-Theorie (2007)

40. Latour, B.: A Relativistic Account of Einstein's Relativity. Soc. Stud. Sci. **18**, 3–44 (1988)
41. Einstein, A.: Relativity: The Special and General Theory, 1st edn. London 1920 (1980)
42. Meadows, D.H., Meadows, D.L., Randers, J., Behrens III, W.W.: The Limits of Growth. New York (1972)
43. Intergovernmental Panel on Climate Change: Summary for Policy Makers. IPCC Special Reports. Emissions Scenarios (2000). ISBN-92-9169-113-5. http://www.ipcc.ch/pdf/special-reports/spm/sres-en.pdf

Chapter 5
The Unknown as an Engine of Science: A Summary

> …as we know, there are known knowns; there are things we know we know. We also know there are known unknowns; that is to say we know there are some things we do not know. But there are also unknown unknowns—the ones we don't know we don't know.[1]

A common theme throughout this essay has been how close major portions of our knowledge lie to the borders of the indefinite. Chapter 2 contained an analysis of the various manifestations of indefiniteness; and in Chap. 3, I have described the value of information as a means to eliminate them. In Chap. 4, "soft" and "hard" methods of dealing with indefiniteness were discussed. It is easy to play with the unknown and use it to lead the public into political adventures as seen in the above excerpt from an interview. Researchers, however, are working with great effort to extend our knowledge beyond the borders of the known, but nevertheless, the boundary between knowledge and the unknown is becoming longer and longer. Rudolf Carnap denied the existence of such a boundary: "Now we can see more precisely what it means that science has no boundaries: every statement based on scientific concepts can be fundamentally determined to be true or false."[2] Does the recognition of the indefinite as a part of our cognitive world contradict Carnap's thesis, which he presented in his book "*Der logische Aufbau der Welt*" (*The Logical Structure of the World*)? Where is the contradiction? Can it be eliminated?

Scientists resolve this contradiction by differentiating between a "remediable" and an "essential" indefiniteness. At the beginning of this treatise, I described six different types of indefiniteness (Sects. 2.1–2.6), characterized as randomness, uncertainty, indeterminacy, vagueness, indistinctness, and undefinedness. In dealing with these very different kinds of indefiniteness, there are, I believe, just two appropriate responses: Depending on whether the indefiniteness can be classified as "remediable" or "essential", a scholar can accept the challenge and intensify his efforts, or he can avoid the risk. I wish now to discuss this topic in some more detail:

[1]Donald Rumsfield, Department of Defense; News Briefing, answer to the question as to whether Iraq has attempted to or is willing to supply terrorists with weapons of mass destruction: http://www.defense.gov/transcripts/transcript.aspx?transcriptid=2636.
[2]Carnap [1].

H.J. Pirner, *The Unknown as an Engine for Science*,
The Frontiers Collection, DOI 10.1007/978-3-319-18509-5_5

In understanding Carnap's thesis, the definition of the word "fundamental" plays an important role. He states: "Concerning the expression 'fundamental': If a question for example about a particular event is practically unanswerable because the event was too distant, spatially or temporally, but a similar question about a current and nearby event is answerable, then we will denote this question as *practically unanswerable* but *fundamentally answerable*".[3] Such fundamentally answerable questions fall in the class of "remediable" indefiniteness. It was shown in Chap. 3 how additional information can reduce or completely eliminate this type of indefiniteness. I suggested information-processing methods, for example the comparison of texts on a large scale to reveal mutual information according to similarities and trends. Supercomputers and Cloud Computing will make this possible. An open reanalysis and investigation of the facts can help eliminate this type of indefiniteness. Complexity does not appear to be an insurmountable obstacle to knowledge, as long as we can isolate the complex internal structure of a system from the indefiniteness of its surroundings. In the physics of condensed matter and in the biological sciences, independent but repeatedly-occurring complex model structures can be identified, for example pair formation by electrons in superconducting materials.

Many uncertainties are due to the fact—as noted by Rudolf Carnap—that the events in question are temporally or spatially too far away from us. The evolution of the early universe 13 billion years ago can be reconstructed in the laboratory only with difficulties, since the temperature and the density of the matter shortly after the Big Bang were extremely high. Nevertheless, high-energy experiments can give us important information about the particles in the universe at these high temperatures. Although the cosmos became transparent for electromagnetic radiation only after some 300,000 years, the minute inhomogeneities in the microwave background radiation observed today can trace quantum fluctuations in the very early universe, which led to the formation of the galaxies. The costs and the constraints of existing technologies limit the maximum energies which can be attained by new particle accelerators; this makes it increasingly hard to investigate still smaller objects, since the energy determines the resolution of the high-energy microscopes. Here, a new technological development, such as laser-plasma acceleration, will be necessary.

In the humanities, the flawed definition of a concept can render a problem opaque or can lead to paradoxes. The ambiguity of language presents a great challenge, since it can allow "impure" conceptual spheres to overlap and mix. Conventional or usage definitions can help here. Science is in this overreaching sense truly without boundaries, since its borders and its limits can be continually pushed outwards.

Not only are the outer boundaries being continually pushed back; but also within a scientific discourse, there can be different perspectives which coexist interchangeably with equal status. In physics, "effective" theories have become established; they describe the empirical phenomena optimally just within a certain size

[3] Carnap [1, p. 254].

(or energy) scale. There has been a long discussion as to whether physical laws explain fundamental properties or emergent properties. It has resolved into the insight that every scale possesses fundamental organizational structures in analogous fashion. Two theories that overlap within a certain range of scales are acceptable if they make the same predictions and if the theory on the larger scale can be derived from the theory on the smaller scale. We can take as a example the elementary gauge theory of the quarks and gluons, which on a scale of $\ll 10^{-13}$ cm contains the fundamental degrees of freedom. On the larger scale of atomic nuclei, from 10^{-13} to 10^{-12} cm, the nucleons (each composed of three quarks) and their interactions through exchange of mesons (quark-antiquark pairs) appear as an expedient description of the physics. The overlapping theory is the description of the mesons and baryons in terms of the quantum chromodynamics of quarks and gluons.

Instead of coexistence and consistency of theories, relativism and constructionism emphasize the arbitrary nature of different perspectives. These two approaches avoid attempting to formulate an overlapping description which would position different perspectives relative to one another. This may represent an intermediate stage of scientific development, one that is characterized by a particular seriousness (dense description), but it can hardly define a future course for science.

Sciences or scholarly disciplines which are normative or governed by laws, such as medicine and jurisprudence, must be forced to accept their mandate for clarification. An indeterminate clinical pattern requires that the medical practitioner find a more precise diagnosis with the aid of higher-resolution methods. This rule becomes controversial only when the more precise examination itself represents a risk to health. Here, the diagnostician needs theoretical help which can consider his ethical goals and the scientific possibilities together and help him to make a rational decision.

Is there also an *essential* indefiniteness in the sciences? If so, does it have anything to do with the limits of our reasoning power and logic? Are there limiting concepts which indicate things that lie beyond the boundaries of the known? Can we comprehend these phenomena in their most general form only with the aid of metaphysics? I believe that the last question cannot be answered within a scientific context. The first question, in contrast, should be answered with "yes". As an example of essential indefiniteness, we have seen the well-known "indeterminacy relations", which do not allow the simultaneous precise determination of position and momentum, for example, in microscopic physics. Similar relations are naturally found in other systems which contain correlated fluctuations, as well. The fluctuations of the (internal) energy and the inverse temperature within a small system which is in contact with a heat bath satisfy a kind of uncertainty relation: $\Delta U \Delta (1/T) \geq k$, where k is the Boltzmann constant.[4] Joos Uffink shows the similarities to and differences from the quantum-mechanical uncertainty relations. The heat bath is

[4]Uffink [2].

the source of the uncertainty in thermodynamics, because the smaller system can exchange energy with the larger system. There is no fundamental source in nature responsible for the thermodynamic uncertainty. "No-go" theorems are found e.g. for black holes, which have the highest entropies of all known physical objects of comparable mass, and which probably contain the greater part of the total entropy of the universe. Black holes within their Schwarzschild radii are closed to our view; we can see only secondary effects, e.g. how the black hole takes on matter from a nearby star. In this process, part of the matter that is being swallowed up is converted into energy and accelerated away in the form of jets of particles or gamma radiation that can be observed.

Science deals with essential indefiniteness by assuming its existence as a hypothesis which must be repeatedly validated and reanalyzed at intervals. These essential cases of indefiniteness are surrounded by speculations which struggle with conceptual uncertainty. In this context, I have mentioned Hugh Everett's hypothesis concerning the existence of "*many worlds*" in quantum mechanics (see Sect. 3.2). The *holographic principle* is another such speculation, which interprets our four-dimensional world as a projection from the surface of a higher-dimensional structure.

In setting itself off from pseudoscience and postmodern arbitrariness, serious scholarship triumphs through its openness. "With the idol of certainty (including that of degrees of imperfect certainty or probability) there falls one of the defenses of obscurantism which bar the way of scientific advance. For the worship of this idol hampers not only the boldness of our questions, but also the rigor and the integrity of our tests. The wrong view of science betrays itself in the craving to be right; for it is not his possession of knowledge, of irrefutable truth, that makes the man of science, but his persistent and recklessly critical quest for truth".[5] This search for definiteness in science is driven by indefiniteness.

In the present essay, I began with the phenomenon of indefiniteness in various forms. I then described the dynamics of how information technology is leading us to a new definiteness. If we can find a basis for considering indefiniteness and definiteness together, only then will we arrive at new scientific knowledge. My main propositions may be summarized as follows:

- Indefiniteness must be investigated in a temporal (past, present, and future) or a conceptual (semantical, theoretical, ontological) context. It can then be classified as random, uncertain, indeterminate, vague, indistinct or undefined. The indefinite of the "environment" and the definite of the "system" must be considered together, but in a differentiated manner.
- Information reduces uncertainty. Conversely—the variation in the complexity of a system relative to the change in the indefiniteness of its environment is a measure of the qualitative value of the information. Empirical distributions or opinion functions can bring precision into indefinite concepts (e.g. rich/poor).

[5]Popper [3].

The internal logic of the concepts is not sufficient. Contextuality in complex networks is the method of choice to narrow down vague limits.

- In signs, the indefinite object of reality merges with the definite phenomenon within a constitutive interrelation. The precursors of this "Platonic" mixture are the assignment, the understanding, and the creation of scenarios. Neither a naive realism, nor a pure phenomenology, nor an entirely theoretical idealism will lead us to firm ground.

- Distinguishing the "practically" inanswerable and the "fundamentally" inanswerable;
- Indefiniteness of the first kind can be eliminated by more information;
- The essential indefiniteness of the "indeterminacy relations";
- The open character of science.

References

1. Carnap, R.: Der logische Aufbau der Welt. Scheinprobleme in der Philosophie, p. 255 (For an English-language commentary, see http://fitelson.org/few/few_05/leitgeb.pdf). Hamburg (1961)
2. Uffink, J.: Found. Phys. **29**(5), 655–692 (1999)
3. Popper, K.R.: Logik der Forschung, Tübingen 1966, p. 225 (The Logic of Scientific Discovery) (1992)

Epilog

At the beginning of the Marsilius project on indefiniteness, I read Alan W. Watts' scholarly introduction to Zen Buddhism.[1] In it, Watts describes an undefined but concrete path to finding meaning with the help of the indefinite. It is the oriental method of "*wu wei*" ("non-doing"), the spontaneous process of allowing the mind to follow its own path. He maintains that this is better than focusing on an object: "The perfect man employs his mind as a mirror. It grasps nothing, it refuses nothing. It receives, but does not keep. One might say that it fuzzes itself a little to compensate for too harsh a reality".[2] He quotes Lao tzu:

(Meaning)… as a thing
Seems indistinct, seems unclear
So unclear, so indistinct
Within it there is image
So indistinct, so unclear
Within it there is substance
So deep, so profound
Within it there is essence
Its essence is supremely real
Within it there is faith
From ancient times to the present
Its name never departs
To observe the source of all things.[3]

These lines of poetry affirm a meaning which does not consider knowledge to be a way of controlling reality, but rather allows the person who has been animated by this meaning to trust in the spontaneous evolution of reality.

The contradiction between the requirement of discipline in interdisciplinary research and accepting spontaneity was thought-provoking. The Kolleg offered a good opportunity to find a common language among the various disciplines and to become acquainted with different scientific cultures in the humanities and the

[1]Cf. Watts [1]. Physics has been interpreted in terms of Taoism by Capra [2]. The book by Vedral [3], p. 218, also closes with a guote from the *Tao te ching*.
[2]Watts [1], p. 39.
[3]Cf. The translation of Tzu [4].

H.J. Pirner, *The Unknown as an Engine for Science*,
The Frontiers Collection, DOI 10.1007/978-3-319-18509-5

sciences. Our knowledge is broken up into individual scholarly disciplines which have become specialized and have expanded enormously within themselves. It is thus necessary today to build bridges and connections between those specialized disciplines. In order to encompass our reality, we need the whole picture of the islands of knowledge in the ocean of indefiniteness. They exist together in our world.

Have we found the right measure for the study of the indefinite? This essay has dealt with methods of surveying indefiniteness. These to some extent new approaches, together with known methods, can facilitate navigation on the ocean of the unknown. Insofar as the indefinite is immeasurable, this attempt can succeed only partially at best; there is simply too much of the indefinite. David Bohm expressed this tension as follows: "But original and creative insight within the whole field of measure is the action of the immeasurable. For when such insight occurs, the source cannot be within ideas already contained in the field of measure but rather has to be in the immeasurable, which contains the essential formative cause of all that happens in the field of measure".[4]

The relationship between the determining and the determined is then no longer epistemological, but rather metaphysical. Genesis 28 recounts how Jacob lies down and dreams: "And he dreamed, and behold a ladder set up on the earth, and the top of it reached to heaven: and behold the angels of God ascending and descending on it. And, behold, the LORD stood above it, and said, I am the LORD God of Abraham…".[5] (The LORD in this version stands for *Adonaj*, the Hebrew name of God. In ancient Hebrew scriptures, the tetragram "jhwh" was used, since in Jewish tradition, God's name is not to be spoken; on reading the text, it is expressed as "haSchem".) Between the regions 'Heaven' (the determining) and 'Earth' (the determined), which are presumed to be known, Jacob's ladder acts as an indefinite connection. The angels move up and down on this ladder, i.e. back and forth between the two worlds. St. Benedict, in his "*Holy Rule*" (*Regula Benedicti*, Chap. 7), interprets this: "By that descent and ascent we must surely understand nothing else than this, that we descend by self-exaltation and ascend by humility".[6] We learn from the Enlightenment that the relationship between the determining and the determined must be personally constructed by each individual. It is in the end the acceptance of the indefiniteness of this construction that admonishes us to practice humility and tolerance.

[4]Bohm [5], p. 50.
[5]Bible [6].
[6]The *Holy Rule of Benedict* [7].

References

1. Watts, A.W.: The Way of Zen. Vintage, New York (1957)
2. Capra, F.: The Tao of Physics: An Exploration of the Parallels Between Modern Physics and Eastern Mysticism, Berkeley (1975)
3. Vedral, V.: Decoding Reality: The Universe as Quantum Information, p. 218. Oxford University Press, Oxford (2010)
4. Tzu, L.: Tao te ching. http://www.taoism.net/ttc/complete.htm
5. Bohm, D.: Die implizite Ordnung. Grundlagen eines dynamischen Holismus (Wholeness and the Implicate Order, Routledge, London), p. 50. Trikont-Verlag, München (1985)
6. Bible, H.: King James Version. https://www.biblegateway.com/passage/?search=Genesis+28&version=KJV
7. The Holy Rule of Benedict (English translation). http://www.osb.org/rb/text/rbejms3.html#7

Appendix

A straightforward example of how complexity and indefiniteness combine to define the value of information—consider the simple function of one variable:

$$F(x) = ax(1 - x).$$

This function can demonstrate in a graphic way how complexity can be produced from something which is extremely simple. Let the control parameter a lie in the interval $1 < a < 4$ and the value range of x be 0–1. Then the graph of this function has the shape of a hat or an upside-down parabola (see Fig. A.1). Its functional values range between 0 and a maximum of 1, where the latter value is reached only when $a = 4$. Many phenomena in physics or biology can be described as repetitive processes. The function F can be used to describe the transition from a state n to the state $n + 1$, that is the mapping of one variable onto a new variable. The iteration then takes on the following form:

$$x_{n+1} = ax_n(1 - x_n).$$

One can ask if the series of numbers $x_1, x_2, x_3, \ldots, x_n$ approaches a well-defined limiting value. This value is called the *fixed point* of the mapping, since it does not change on further iteration. It is determined by the equation

$$x_\infty = ax_\infty(1 - x_\infty).$$

and can be readily computed: $x_\infty = (a - 1)/a$ for $a > 1$.

In order to test the stability of the fixed point, one adds a small perturbation δ to this expression. On iteration, δ generates a deviation of $\delta(2 - a)$. To prevent this deviation from increasing, its absolute value $|2 - a|$ must be <1, or $1 < a < 3$. Such a fixed point is an *attractor*, i.e. it attracts every starting value to this limiting functional value after a certain number of iterations. In Fig. A.2, we can see[7] that a few iterations suffice to arrive at the fixed point $x_\infty = 0.5$ for a control parameter

[7]Weisstein [1] lists a program which calculates and plots the values; Figs. A.2 and A.3 are taken from this site.

H.J. Pirner, *The Unknown as an Engine for Science*,
The Frontiers Collection, DOI 10.1007/978-3-319-18509-5

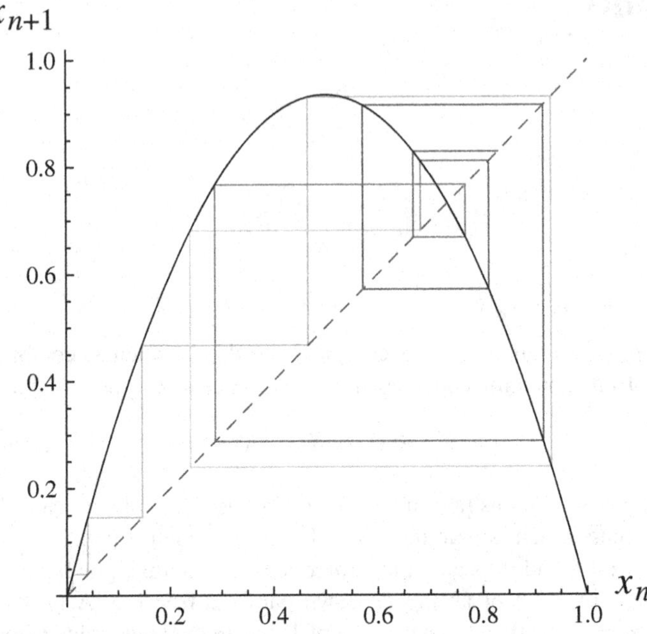

Fig. A.1 A graphic representation of the logistic mapping: The values of x_n are plotted on the horizontal axis, and the corresponding values of x_{n+1} (the functional values) on the vertical axis. The iteration can be traced by starting from the given x_n-value and going vertically to the *black parabola*, then horizontally to the dashed line, which gives the new starting value for the next step. As soon as one reaches the *right-hand side* of the parabola, the horizontal steps produce a large change in the value of x

$a = 2$, independently of the starting value of x. Increasing a to 3 produces an oscillatory limiting behavior between the functional values 0.66 and 0.67. This twofold solution is replaced by a fourfold solution (0.5, 0.88, 0.38, 0.83) for $a = 3.5$.

The splitting up into more and more limiting functional values is termed *bifurcation*. A so-called *Feigenbaum diagram* plots the limiting values against the central parameter a and shows how they split up repeatedly and finally become a dense series of rapidly varying values for a greater than the Feigenbaum number, $a = 3.5699$. With $a = 4$, even an infinitesimal variation of the starting value from x_1 to x'_1 causes dramatic changes in the functional value after 10–15 iterations. This is the *chaotic* region of the mapping, in which the starting values x_1 or x'_1 lead after n iterations to final values x_n or x'_n whose difference grows exponentially

$$|x_n - x'_n| = \exp(n\lambda) \quad (\lambda > 0).$$

The positive *Lyapunov exponent* λ describes how strongly the nth iteration values differ for a given pair of initial values. The behavior of the solutions of the logistic

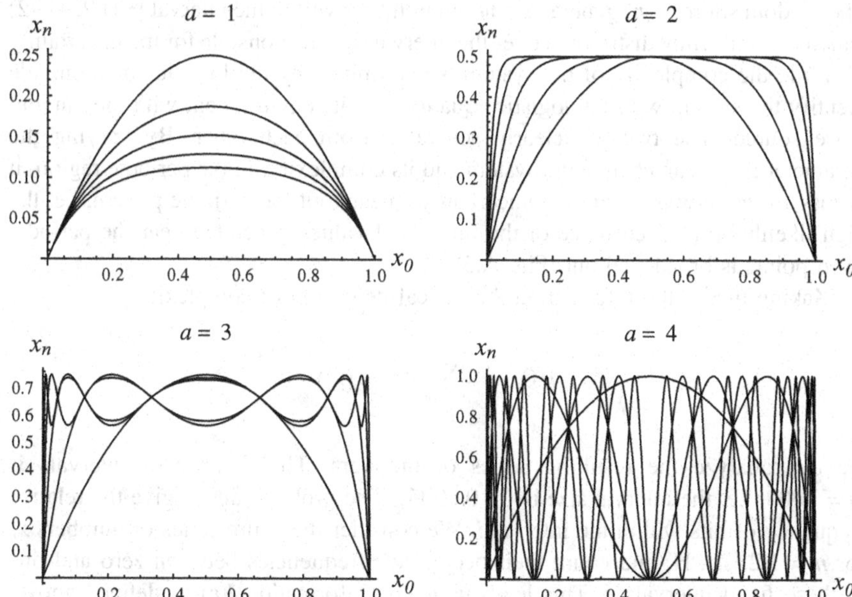

Fig. A.2 The dependence of the iteration values on the initial value of x. The control parameter a used is shown in Fig. A.1. After one iteration, a point on the parabola is found for every value of a. For $a = 1$, the values from successive iterations become smaller and smaller. With $a = 2$, one recognizes the fixed point at 0.5. For $a \geq 3$, the bifurcation into two alternating limiting values can be seen; for $a = 4$, the chaotic region has been reached

equation are similar to the behavior of the trajectories of Newtonian equations of motion with many degrees of freedom, whose mechanical motion is chaotic. When the initial values are made 100 times more precise, the equation above, with a Lyapunov exponent of $\lambda = 1$, gives only a small improvement in predictive power, i.e. n can be increased only to $n + 4$. For a mechanical system, n is proportional to the time, and an improved precision of the initial values gains us only a minimally longer prediction time. Put differently, for all chaotic systems we can predict the future with only very limited success. Within the chaotic region itself, there are again ordered subregions with periodic behavior and negative Lyapunov exponents.

We will not discuss this in detail here, but instead refer the reader to the literature.[8]

In the following, we will study such a deterministic system under the influence of an external source of random disturbance, which introduces noise into the system:

$$x_{n+1} = ax_n(1 - x_n) + \nu(n).$$

---

[8]Goldstein et al. [2], p. 523 ff.

The random source $v(n)$ generates random numbers within the interval [−U/2, +U/2] which are uniformly distributed over the interval. It is responsible for the uncertainty U, while the complexity of the system is determined by the logistic equation. We identify the system with the logistic equation and its environment with the random noise source. The two are clearly separated from each other. By varying the parameter a, we can change the system and its complexity. In the periodic region, it is easy to see how an increased uncertainty smears out the periodic pattern, i.e. the high likelihood of occurrence of the functional values generated near the periodic fixed points is broadened out (Fig. A.3).

Making use of the information-theoretical definition of complexity:

$$I(\varepsilon) = -\sum_{i=1}^{N} p_i \log_2(p_i)$$

we can analyze the resulting series of numbers. The number of intervals is N = 2000, i.e. the interval size is (0, 0.0005). The probabilities p_i give the relative frequency of hits within the interval i. We consider the entire series of numbers x_n for n = 1,2, …, N, and count their occurrence frequencies between zero and one in these fixed intervals ε. This leads us to the information $I(\varepsilon)$ as defined above. We carried out such an "experiment" in the neighborhood of the boundary to chaos, i.e. at the Feigenbaum number a = 3.599. It showed that the complexity of the

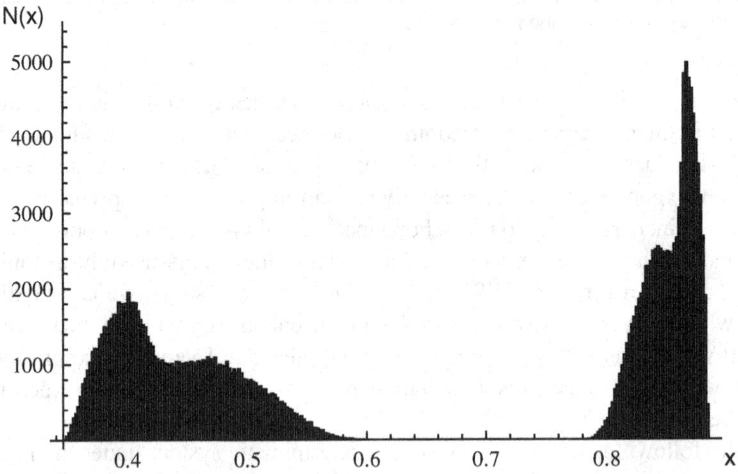

Fig. A.3 For an uncertainty of U = 0.02, the periodic pattern of functional values of the logistic equation with a = 3.5, which has the fourfold fixed points 0.382, 0.501, 0.827, and 0.875, is broadened out. The resulting complexity is determined from the information given by the number of events (values) within discrete intervals of width ε = 0.0005. The change of complexity as a function of the uncertainty for different values of the control parameter a and the uncertainty parameter U then gives the value WIN

sequence of numbers increases as expected with the uncertainty, exhibiting an approximately linear dependence of the complexity on the logarithm of the uncertainty:

$$\mathrm{WIN} = \Delta \mathrm{K}(\varepsilon)/\Delta \mathrm{Log}[\mathrm{U}]$$

The further into the chaotic region we progress, the richer—i.e. the higher its information content—is the resulting pattern of numbers. The specific increase in information becomes smaller as a increases from $a = 3.5$ to $a = 3.8$, which can be seen simply by inspection. The ratio WIN changes from WIN(periodic) = 0.87 to WIN(chaotic) = 0.04, and it reflects the weaker increase of the complexity with equal changes in the uncertainty at higher values of a. In the chaotic region, the system itself is already rather complex, so that adding an external uncertainty causes little change in the complexity of the system. This means that the gain in genuinely valuable information is smaller in the chaotic region.

The degree of dynamic information can also be considered as a function of time by identifying the length of the number sequence N with the time $T = \mathrm{N}\tau$, where each step in time has the length τ. For deterministic chaotic processes like the logistic equation, the Kolmogorov-Sinai definition of complexity is useful. The algorithmic complexity is given by the length of the shortest program which would reproduce the number sequence generated by the equation on a universal Turing machine. A Turing machine is an idealized model of a computer, with a read-write device and a linear program tape. The symbol | .. | defines the binary length of the program. We first consider the periodic region of the logistic equation, in which the equation produces the same sequence of numbers again and again. The associated program then contains only the periodically-occurring fixed points and a command that tells how often the sequence is to be repeated. Coding the number of repetitions $\mathrm{N} = T/\tau$ requires log[N] bits. Along with programming the number N, one requires a finite number of additional bits which specify the periodic numbers, but they add little to the program length. For long times, the Kolmogorov-Sinai complexity therefore grows logarithmically with time, $\mathrm{KS} \approx \log[T]$. However, in the chaotic region, all N numbers of the sequence have to be specified by the program; its length then grows proportionally to the time, $\mathrm{KS} \approx T$. It can be shown how the Kolmogorov-Sinai entropy is related to the Lyapunov exponent.[9]

When noise is added to the logistic equation, the resulting complexity depends on the random signal from the noise generator, thus also on the time T. For such random processes, there is a generalization of the K-S complexity termed the ε entropy, which analyzes the increase in the entropy as a function of the time T and the size of the intervals, ε.[10] When the uncertainty from the noise generator is zero, the ε entropy remains constant per unit time. If the resolution $\varepsilon < U$ is smaller than the uncertainty, the specific entropy per unit time increases as $\log[U\varepsilon T]$; it saturates

[9]Cf. Falcioni et al. [3].

[10]Cf. Gaspard and Wang [4], pp. 291–343.

when the resolution becomes larger than the fluctuations of the uncertainty signal. An extended analysis of the time sequence of discrete, randomly-modified logistic mappings by Pierre Gaspard and Xiao-Jing Wang[11] showed that the complexity per time step increases with the uncertainty.

- The complex system is represented by the logistic equation;
- Its environment is simulated by a random term (uncertainty);
- The resulting number sequence represents the information or complexity;
- This depends in a controlled fashion on the uncertainty.

References

1. Weisstein, E.W.: MathWorld. http://mathworld.wolfram.com/LogisticMap.html
2. Goldstein, H., Poole, C.P., Safko, J.L.: Klassische Mechanik, p. 523 ff. Wiley, Weinheim (2006)
3. Falcioni, M., Loreto, V., Vulpiani, A.: Kolmogorov's legacy about entropy, chaos and complexity. In: The Kolmogorov Legacy in Physics. Lecture Notes in Physics 636, pp. 85–108. Springer, Heidelberg (2003)
4. Gaspard, P., Wang, X.J.: Noise, chaos, and (ε, τ)-entropy per unit time. Physics Report **235**(6), 291–343 (1993)

[11]Gaspard and Wang [4], p. 313.

Glossary

Age of the universe The *age* of the Solar System can be estimated from the decay of naturally-occurring radioactive isotopes, and is found to be about 4.6 billion years. Based on calculations of the relative abundances of the elements following supernova explosions, the age of the Milky Way galaxy is estimated to be 11 billion years. Assuming that the universe began with the Big Bang, and making use of the cosmic microwave background and the Standard Model of cosmology, its age is computed to be 13.8 billion years. One billion (10^9) years is 1,000,000,000 years. (The notation 10^n means a 1 followed by n zeroes, e.g. $10^2 = 100$.) The characteristic time scale for the expansion of the universe is the Hubble time, $\tau = 9.8 \times 10^9\ h^{-1}$ years, with $h \approx 0.72 \pm 0.03$ (the Hubble constant)

Anthropic principle In cosmology and philosophy, the *anthropic principle* denotes the argument that the physical universe must be compatible with the existence of human life. This principle limits the possible initial conditions for the formation of the universe

Atomic weight The (relative) *atomic weight* of an element is the mass of one of its atoms relative to the mass of carbon-12, taken to be exactly 12 atomic mass units (amu). The nucleus of a carbon-12 atom contains 6 protons and 6 neutrons. Since protons and neutrons (the *nucleons*) have only slightly differing masses, the relative atomic weight gives the number of nucleons within an atomic nucleus

Black holes Regions in spacetime from which no light can escape to the outside world

Bose (-Einstein) condensate A *Bose-Einstein condensate* is the ground state of an ensemble of many indistinguishable bosons (particles with integral spin) at extremely low temperatures. At higher temperatures, the bosons have differing momenta, but at low temperatures, many of the bosons have zero momentum and thus zero kinetic energy these form the condensate

Boson *Bosons* are particles with integer spin (proper angular momentum) quantum numbers, $J = 0, 1, 2, \ldots$. In the Standard Model of elementary particles, bosons are the elementary particles which *mediate* the fundamental interactions, i.e. they

H.J. Pirner, *The Unknown as an Engine for Science*,
The Frontiers Collection, DOI 10.1007/978-3-319-18509-5

transmit the basic forces (electromagnetic, weak, strong) between the matter particles (fermions)

Complex numbers *Complex numbers* were introduced in order to define numbers whose squares are negative. One can represent a complex number as a point on a (two-dimensional) plane. Its distance to the coordinate origin in the plane is the *magnitude* of the number, and the angle it makes with the x axis is the *phase* of the complex number. Alternatively, the two coordinates x (real part) and y (imaginary part) can be specified

Correlations *Correlations* $\langle (X - \langle X \rangle)(Y - \langle Y \rangle) \rangle$ indicate to what extent the deviations of a quantity X from its mean value $\langle X \rangle$ are related to (correlated with) the deviations of a quantity Y from its mean value $\langle Y \rangle$

Critical point In the theory of phase transitions (e.g. gaseous ↔ liquid, or spontaneously magnetized ↔ unmagnetized), a *critical point* denotes the end point of a phase-boundary line, for example in a pressure-temperature diagram (phase diagram). At the critical point, the densities of the gaseous and the liquid phases approach each other

Dark energy *Dark energy* dominates the overall energy density of the universe, accounting for a fraction of 69 % of the total energy density. If this component were constant in time, it would obey an equation of state w = pressure/energy density = −1 and can be described by a cosmological constant

Dark matter The existence of matter which neither emits nor absorbs light—thus "*dark matter*"—is held to be well established. The best evidence for its existence comes from observations of stars, gas clouds and clusters whose rotational motion (e.g. within galaxies) is more rapid than can be explained by the density of visible matter around them. Dark matter accounts for about 26.2 % of the overall energy density of the universe; what it consists of is unknown

Decoupling In the thermal evolution of the expanding universe, the time at which it became transparent to light is particularly important. This time limits our ability to look back in time to the early universe. As the universe cooled, electrons could be captured by protons to form hydrogen atoms. When this process was complete, photons (light particles) could no longer encounter free electrons and be scattered by them. They do not scatter from the neutral hydrogen atoms, which consist of a positive proton and a negative electron. One speaks of the *decoupling* of the photons; the universe became transparent to electromagnetic radiation at this time

Duality Different theories which correspond to each other in certain limits are said to be *dual* to one another. The best-known example is a system of elementary spins on a square lattice in two dimensions. Each spin has the possible orientations +1 and −1, and they interact with each of their four nearest neighbors. The dual system consists of elementary squares whose values are given by the product of the combined orientations of neighboring spins

Electron volt One *electron volt* (eV) is the energy that a particle with one elementary charge e_0 (the electric charge of an *electron*) gains when it moves through an electric potential difference of *one volt*

Emergence *Emergence* refers to the formation of complex systems from simple elementary components, whose individual properties alone cannot explain the properties of the whole system. The whole is more than the sum of its parts, or "more is different" (Philip W. Anderson)

Energy density of the universe The overall *energy density* of the universe (total energy/volume) is the sum of the densities of ordinary matter (baryonic matter, 4.8 %), radiation (0.001 %), dark matter (26.2 %) and dark energy (69 %). Its value is fixed within narrow limits by the observed flat geometry of the universe

Entropy *Entropy* is a measure of the disorder in a physical/chemical system. For example, if we have M different states which are all found with the same probability, the entropy of the system is given by $S = k \log M$. In a pure state, $M = 1$ and the entropy is $S = 0$ (no disorder). The factor k is Boltzmann's constant, a physical constant which relates energy to temperature

Fermion *Fermions* are particles with half-integer spin (proper angular momentum) quantum numbers, $J = 1/2, 3/2, 5/2, \ldots$. In the Standard Model of elementary particles, the fermions are *matter particles*

Fourier transformation Fourier transformations are integral transformations which relate a given function $f(x)$ to a function $F(p)$ via integration with an exponential function: $F(p) = \int f(x)\, \mathrm{e}^{ipx}\, \mathrm{d}x$. The inverse transformation is $f(x) = (1/2\pi) \int F(p)\, \mathrm{e}^{-ipx}\, \mathrm{d}p$

Gaussian distribution The *Gaussian* or *normal distribution* is a probability distribution which has the form of a bell curve. Its importance is due to the *central limit theorem*, which states that the mean values of n independent random observations take the form of a Gaussian distribution as n becomes large

Hawking radiation Fluctuations of the vacuum, producing electron-positron pairs in the neighborhood of a Black Hole, can lead to absorption of one of the particles into the hole and the emission ("evaporation") of the other. The energy of the emitted particle comes from the gain in potential energy of the particle that is absorbed (falls into the hole). This leads to a net energy current out of the Black Hole. The smaller the Black Hole, the stronger the curvature of spacetime in its neighborhood and thus the stronger the evaporation of particles. The energy current away from the Black Hole is termed "*Hawking radiation*"

Hebb's rule *Hebb's rule* states that the change $\Delta w(i)$ of the weight of the ith synapse is proportional to the associated input potential $x(i)$ and to the activation function $y(a)$. It is used to explain associative learning

Higgs boson In the Standard Model of elementary particles, the scalar (spin $J = 0$) *Higgs field* is responsible for the masses of all the elementary particles. The fluctuations of the Higgs field around its vacuum expectation value are called the *Higgs boson*; it is named after the British theoretical physicist Peter Higgs, who shared the Nobel Prize for physics in 2013 with Francois Englert. The Higgs Boson was discovered in 2012 at CERN, Geneva, at a mass corresponding to 125 GeV

Hooke's law In classical mechanics, *Hooke's law* states that the extension of a perfectly elastic spring is directly proportional to the force which is acting on it

Hydrodynamics The field of *hydrodynamics* studies the motion of fluids

Inertial system An *inertial system* or *inertial frame* is a coordinate system that moves relative to other frames of reference with constant velocity (constant speed and in a straight line, i.e. without acceleration)

Inflation of the universe (Cosmic) *inflation* refers to an extremely strong and rapid expansion of the universe shortly after the Big Bang, i.e. in its earliest stages

Information (Syntactic) *information* as defined by Claude Elwood Shannon is given by the degree of randomness of a string of characters; the more improbable a character, the higher the information content that it conveys to the recipient

Initial conditions The data which describe the initial state (at time $t = 0$) of a physical system (e.g. the positions and velocities of the particles)

Isotopes *Isotopes* are atoms of a particular element which have different atomic weights. Their atomic nuclei contain the same numbers of protons (same nuclear charge and thus the same chemical element), but differing numbers of neutrons (and thus different atomic weights)

Kelvin The temperature scale which begins at absolute zero, 0 K = −273.15 °C. The unit "one Kelvin" is the same as "one degree Celsius", but the value "0" on the Kelvin temperature scale is shifted (by 273.15°) to lower temperature from that of the Celsius scale

Light cone The *light cone* that belongs to a particular point in spacetime is the geometric locus of all light trajectories which could originate at that point

Logarithm The *logarithm* is defined in elementary mathematics as a function which permits multiplication to be represented in terms of addition: $\log(x \cdot y) = \log(x) + \log(y)$. It is often evaluated with the aid of logarithm tables or a slide rule. The perceived loudness of a sound is proportional to the logarithm of its intensity (the sound power per unit area)

Log-Normal distribution In probability theory, a log-normal distribution is the result of the products of random quantities which are all positive. A log-normal distribution of X means that log(X) follows a normal distribution (bell curve)

Matrix (plural: matrices) A matrix is a rectangular table of elements, organized in rows and columns. Matrices can be added, multiplied, etc

Mean square deviation The *mean square deviation* σ is defined as the square root of the expectation value of the squared deviation from the mean value of a set of values X: $\sigma = \sqrt{\langle (X - \langle X \rangle)^2 \rangle}$

Mean value The *mean value* of a set of randomly-distributed values X with the probability distribution $p(X)$ is given by the integral $\langle X \rangle = \int Xp(X)\, \mathrm{d}X$

Microwave cosmic background radiation The *cosmic microwave background* is the radiation left over from the earlier, hotter universe; it was present at the time when the electrons and protons combined to form atoms (decoupling). Due to the expansion of the universe in the intervening time, this radiation is now detectable in the microwave region (wavelengths of a few centimeters), corresponding to a radiation temperature of only 2.725 K

Newton's laws Isaac Newton formulated three *laws of motion*, the law of *inertia*, the law of *action*, and the law of *reaction* (*actio* = *reactio*). The law of action is often called simply "*Newton's law*"; it states that the acceleration a of a point mass is proportional to the force F acting upon it, $F = ma$ (m = inertial mass)

Phase relation Quantum-mechanical wavefunctions contain complex terms, each of which consists of a magnitude (amplitude) and a *phase*. Amplitudes with the same phase can reinforce one another, while amplitudes of opposite phase cancel out

Phase space In classical mechanics, a point mass is uniquely characterized by its velocity and its position (three spatial components each). The set of all such possible triplets of velocity and position for N particles defines a ($6N$-dimensional) space, called *phase space*

Phase transition A physical or chemical system can occur in various forms, termed *phases* (e.g. solid, liquid, gaseous). A *phase transition* is the crossing of the system from one phase to another (e.g. melting, solidification, evaporation, condensation, sublimation); i.e. crossing over a *phase boundary*

Physical constants *Physical constants* are quantities which are independent of space and time. They include for example the velocity of light, Planck's constant, the elementary electric charge and the masses of electrons and of protons

Poisson distribution Chance events such as the radioactive decay of atomic nuclei are well described by a Poisson distribution. Starting with a large number N of radioactive nuclei, with the probability w for decay of a particular nucleus per unit time (decay constant), we find in a short time interval $\mathrm{d}t$ an average number $m = N \cdot w\,\mathrm{d}t$ of decays. The probability P of observing k decays within this interval is then given by the Poisson distribution as $P(k) = m^k/k!\ \mathrm{e}^{-m}$

Probability calculus *Probability calculus* yields the probabilities that chance events will occur or will not occur. In order to obtain the probability of a particular chance event, one counts how often such an event occurs and divides that number by the number of all possible events in the same time period

Quantum field theory *Quantum field theory* is the combination of quantum mechanics with field theory. A prime example is quantum electrodynamics, the theory of electromagnetic radiation and its interactions with matter

Quantum mechanics *Quantum mechanics* (or, more generally, *quantum theory*) is a theory of the microscopic structure of matter and the interactions of microscopic particles which is based on Planck's constant as the fundamental unit of quanta. Quanta are the indivisible units in which energy, angular momentum and other quantities are created or annihilated, or change form

Relativity theory One distinguishes between the *special* and the *general theories of relativity*. *Special* (or classical) relativity starts from the speed of light as limiting velocity and derives the transformation formulas for various physical quantities between different inertial frames; it shows that time and space are not absolute. *General* relativity (Einstein, Hilbert) explains gravitation as a result of the curvature of four-dimensional spacetime

Sphere (*n*-dimensional) The n-dimensional volume of an n-dimensional sphere (hypersphere) of radius r is $r^n\,c(n)$. Its $(n-1)$-dimensional surface area is found by taking the derivative of the volume with respect to the radius, $nr^{n-1}c(n)$. A spherical shell of thickness Δr has the volume $nr^{n-1}\Delta r\, c(n)$

Spin system A *spin system* is a set of particles which have proper or internal angular momentum ("*spin*"), and which can interact with each other on the basis of this angular momentum. Permanent magnets owe their magnetism to the interactions of such elementary angular momenta, which tend to orient in the same direction, i.e. they are spontaneously magnetized

Standard Model The *Standard Model* of elementary particles contains matter particles, the Higgs field, and force-mediating particles. The latter transmit the electromagnetic, weak and strong forces between the matter particles. They include the *photon*, the heavy *vector bosons*, and the *gluons* (all *bosons*, with spin = 1). The matter particles (*fermions*) occur in three families of increasing mass; each family contains very light *neutrinos* (which are electrically neutral), charged *leptons* (electrons, electric charge e_0), and *quarks* (with fractional electric charges of $\pm 1/3\ e_0$ or $\pm 2/3\ e_0$). Note that there is also a Standard Model of astrophysics, which deals with the structure and evolution of stars, including the sun; and a Standard Model of cosmology, which treats the origin, structure and evolution of the cosmos (universe)

Statistical ensemble A *statistical ensemble* contains various groups of many-body systems which all have the same macroscopically measurable properties, but differ in their microscopic structures

String *Strings* are the fundamental, one-dimensional threadlike units of string theory. Vibrations or excitations of these units may lead to all the elementary particles (e.g. those in the Standard Model)

String theory *String theory* asserts that the elementary particles are not pointlike objects, but rather 1-dimensional oscillating threads, which only appear to be pointlike because of the limited resolution of all of our observation methods to date. The theory attempts to combine quantum mechanics and general relativity into a consistent unified theory

Superfluid *Superfluids* avoid the classical phase transition from the liquid to the solid phase at low temperatures. They are formed through (Bose-Einstein) condensation of atoms or pairs of atoms (depending on whether the atoms are themselves bosons or fermions), and are characterized by frictionless flow (vanishing viscosity)

Thermodynamic equilibrium A system is in *thermodynamic equilibrium* with another system when, through exchange of energy (heat) over time, the two systems have arrived at the same temperature (which then no longer changes; the net energy flow between the two systems is then also zero)

Trajectory of a point mass In classical mechanics, every particle is assigned spatial coordinates which define its position in space. The sequence of these coordinates as a function of time (when the particle is moving) is then referred to as its *trajectory*

Unit of measure The numerical values of physical quantities are given in terms of *units of measure*, which are agreed upon by convention. This allows various measurements made at different locations to be compared directly. For example, lengths are given in *meters* (or multiples of meters), masses in *kilograms* (or multiples of kilograms) in the International System of Units (SI)

Vacuum A *vacuum* consists ideally of completely empty space, devoid of any matter, and can be approached by pumping the air out of a container (vacuum chamber). In quantum field theory, *the vacuum* is the state of the universe without particles; it is however difficult to achieve, since quantum fluctuations create particle/antiparticle pairs spontaneously. In cosmology, the associated energy is much greater than the dark energy that is required by the observed accelerated expansion of the universe

Titles in this Series

Quantum Mechanics and Gravity
By Mendel Sachs

Quantum-Classical Correspondence
Dynamical Quantization and the Classical Limit
By Dr. A.O. Bolivar

Knowledge and the World: Challenges Beyond the Science Wars
Ed. by M. Carrier, J. Roggenhofer, G. Küppers and P. Blanchard

Quantum-Classical Analogies
By Daniela Dragoman and Mircea Dragoman

Life—As a Matter of Fat
The Emerging Science of Lipidomics
By Ole G. Mouritsen

Quo Vadis Quantum Mechanics?
Ed. by Avshalom C. Elitzur, Shahar Dolev and Nancy Kolenda

Information and Its Role in Nature
By Juan G. Roederer

Extreme Events in Nature and Society
Ed. by Sergio Albeverio, Volker Jentsch and Holger Kantz

The Thermodynamic Machinery of Life
By Michal Kurzynski

Weak Links
The Universal Key to the Stability of Networks and Complex Systems
By Csermely Peter

The Emerging Physics of Consciousness
Ed. by Jack A. Tuszynski

H.J. Pirner, *The Unknown as an Engine for Science*,
The Frontiers Collection, DOI 10.1007/978-3-319-18509-5

Quantum Mechanics at the Crossroads
New Perspectives from History, Philosophy and Physics
Ed. by James Evans and Alan S. Thorndike

Mind, Matter and the Implicate Order
By Paavo T.I. Pylkkanen

Particle Metaphysics
A Critical Account of Subatomic Reality
By Brigitte Falkenburg

The Physical Basis of the Direction of Time
By H. Dieter Zeh

Asymmetry: The Foundation of Information
By Scott J. Muller

Decoherence and the Quantum-To-Classical Transition
By Maximilian A. Schlosshauer

The Nonlinear Universe
Chaos, Emergence, Life
By Alwyn C. Scott

Quantum Superposition
Counterintuitive Consequences of Coherence, Entanglement, and Interference
By Mark P. Silverman

Symmetry Rules
How Science and Nature are Founded on Symmetry
By Joseph Rosen

Mind, Matter and Quantum Mechanics
By Henry P. Stapp

Entanglement, Information, and the Interpretation of Quantum Mechanics
By Gregg Jaeger

Relativity and the Nature of Spacetime
By Vesselin Petkov

The Biological Evolution of Religious Mind and Behavior
Ed. by Eckart Voland and Wulf Schiefenhövel

Homo Novus–A Human without Illusions
Ed. by Ulrich J. Frey, Charlotte Störmer and Kai P. Willfiihr

Brain-Computer Interfaces
Revolutionizing Human-Computer Interaction
Ed. by Bernhard Graimann, Brendan Allison and Gert Pfurtscheller

Extreme States of Matter
On Earth and in the Cosmos
By Vladimir E. Fortov

Searching for Extraterrestrial Intelligence
SETI Past, Present, and Future
Ed. by H. Paul Shuch

Essential Building Blocks of Human Nature
Ed. by Ulrich J. Frey, Charlotte Störmer and Kai P. Willführ

Mindful Universe
Quantum Mechanics and the Participating Observer
By Henry P. Stapp

Principles of Evolution
From the Planck Epoch to Complex Multicellular Life
Ed. by Hildegard Meyer-Ortmanns and Stefan Thurner

The Second Law of Economics
Energy, Entropy, and the Origins of Wealth
By Reiner Köummel

States of Consciousness
Experimental Insights into Meditation, Waking, Sleep and Dreams
Ed. by Dean Cvetkovic and Irena Cosic

Elegance and Enigma
The Quantum InterviewsThe Quantum Interviews
Ed. by Maximilian Schlosshauer

Humans on Earth
From Origins to Possible Futures
By Filipe Duarte Santos

Evolution 2.0
Implications of Darwinism in Philosophy and the Social and Natural Sciences
Ed. by Martin Brinkworth and Friedel Weinert

Probability in Physics
Ed. by Yemima Ben-Menahem and Meir Hemmo

Chips 2020
A Guide to the Future of Nanoelectronics
Ed. by Bernd Hoefflinger

From the Web to the Grid and Beyond
Computing Paradigms Driven by High-Energy Physics
Ed. by Rene Brun, Federico Carminati and Giuliana Galli Carminati

Printed in the United States
By Bookmasters